Thunderstorms and Airplanes

Also by

Richard L. Collins

FLYING SAFELY

FLYING IFR

FLYING THE WEATHER MAP

TIPS TO FLY BY

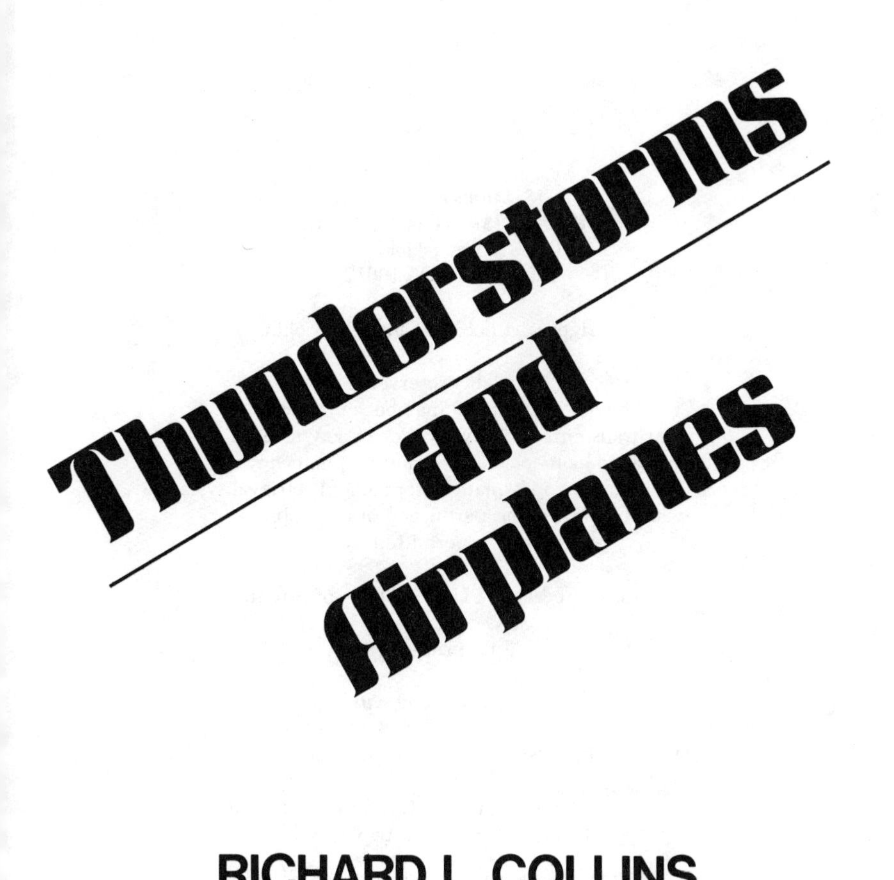

RICHARD L. COLLINS

DELACORTE PRESS/ELEANOR FRIEDE

Published by
Delacorte Press/Eleanor Friede
1 Dag Hammarskjold Plaza
New York, N.Y. 10017

Manufactured in the United States of America

First printing

Designed by Oksana Kushnir

LIBRARY OF CONGRESS CATALOGING IN PUBLICATION DATA

Collins, Richard L., 1933–
Thunderstorms and Airplanes

Includes index.
1. Thunderstorms. 2. Meteorology in aeronautics.
3. Airplanes—Piloting. I. Title.
TL557.S7C64 629.132'5214 81-17533
ISBN 0-440-08877-1 AACR2

CONTENTS

FOREWORD

The relationship of thunderstorms to airplanes is so important that pilots (and passengers, too) have a strong interest in these marvels of nature.

A couple of years ago I got into an interesting discussion of thunderstorms with the administrative assistant of a United States congressman. He was concerned about airline accidents in, around, or under thunderstorms. As folks in the legislating business are wont to do, he thought a new law was needed to put a stop to this. Make it against the law to fly in or near a thunderstorm. Or shut down an airport when a thunderstorm is nearby. He even thought of requiring airlines to post notice at the gate when a departing aircraft would be operating in or near an area where thunderstorms were forecast. This would warn the passenger that some additional risk might be involved in the flight and offer the option of waiting for a better day.

I tried to explain the dynamics of thunderstorms—how they develop and dissipate and how a route might be a maelstrom one minute and reasonably calm the next. Or vice versa. My point was that storms defy precise definition. They are like sneezes. When you have a cold or hay fever, conditions exist for sneezing, but who can pinpoint in advance the precise time and place the sneeze will occur?

I argued that, as a practical matter, the judgment on thunderstorms was something that had to be left to the pilot. Only the view and feel from the left front seat of an airplane can tell the tale. Or in preflight, only the knowledge of what makes thunderstorms, and their relationship to airplanes, can equip the pilot to make the good decisions that lead to a safe flight.

Thunderstorms prey on the minds of all pilots. Much research is done on the subject. The end result is always an admonition to stay out of storms, but staying out of trouble takes more than a resolve. It takes savvy.

Understanding the basics is the first step. Learning from the researchers is another. Accidents are the grade for those who tried and failed, and they offer the ultimate lessons. The aim of this book is not to make anyone feel that he can go out and fly through thunderstorms, but it will provide an understanding of what they are, what they do, how they affect airplanes, and what we can do to avoid them. There's no way to legislate safety in storms; in the final analysis, thunderstorms are best handled by good judgment on the part of the pilot.

RICHARD L. COLLINS

I THE BASICS

The ingredients of a thunderstorm are well known. Lifting action, instability, and moisture are all necessary to get one going. We learn from textbooks that lifting comes from fronts, heating, convergence, or wind flow over terrain. Stability, or instability, is a reflection of the properties of a vertical section of air. Temperature is a key. The air is said to be unstable when it cools more rapidly than 2°C per 1,000 feet. This instability means the air will react enthusiastically to lifting.

Of the necessary things, moisture is perhaps the easiest of the three for a pilot to spot. It can be seen in the form of clouds and reduced visibility, and it can be felt on a hot and sticky day.

It is relatively easy to recognize some of these things from home. You can feel the humidity, and if heat is the lifting agent, it, too, is rather obvious. If the other ingredients are around, the National Weather Service is good at predicting the chance of thunderstorms, and their thundery arrival is usually slow enough to give plenty of time for folding the lawn furniture and rolling up the car windows.

It's different in an airplane. The speed of the machine means we can go quickly from good smooth air into a thunderstorm area. It's not a matter of listening for thunder and then taking some action when it gets close. Pilots have to think 50 or 100 miles ahead, and where weather systems move relatively slowly over ground positions, an airplane can move through systems rapidly. To stay out of the beasts, a pilot has to judge the thunderstorm potential accurately as the situation changes.

We usually think in terms of two basic kinds of thunderstorms: air mass and frontal. The first type develops within an air mass, meaning there is no front pushing through. Lifting is usually generated by heat, though it can also come from wind flowing over mountains. If there is an advantage to air mass storms, it is in their usually being scattered, in relatively isolated clusters or in clearly visible broken lines along a mountain range. Storms that form in squall lines, in widespread clusters, or that are embedded in other clouds, are usually associated with low-pressure systems and associated fronts even though they are not necessarily related to the surface position of a front. The key to understanding the potential of all storms is a basic understanding of circulation patterns at the surface and aloft. The relationship between these circulations can produce instability by positioning cooler air over warm, humid air at the surface. The front or the circulation into the low-pressure center can provide lifting.

Figure 1 shows a low-pressure area with a cold and warm front running off in the usual directions. The low is over Chicago.

Air flows counterclockwise around a low in the northern hemisphere, and tends to assume the properties of the area over which it flows. Warm air will hold more moisture than cold (for every rise in temperature of 11°C, the capacity of air to hold water vapor is doubled). The characteristics of fronts will be discussed shortly (pp. 19–36).

The flow and the characteristics of air in Figure 1 show the southeastern United States under the influ-

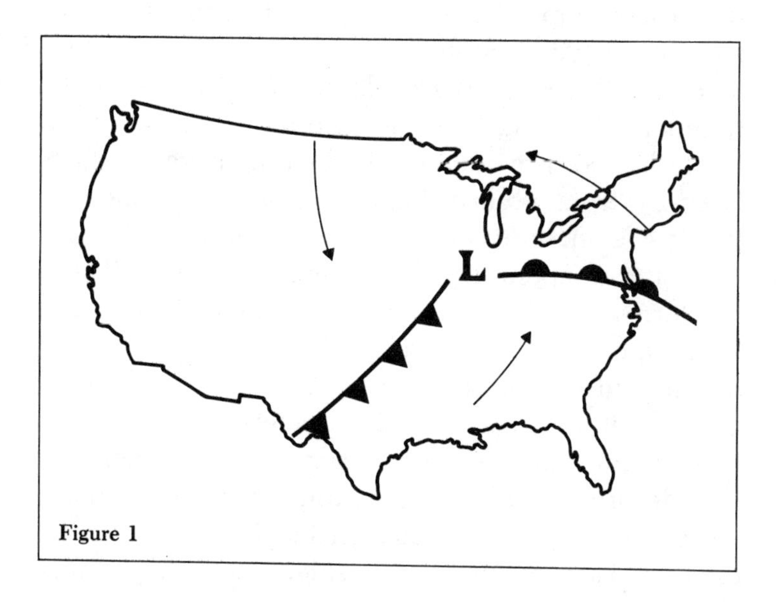

Figure 1

ence of warm, moist air. Warm because it is being circulated from the south and moist because it is being drawn from over the Gulf of Mexico. The fact that it is warm means it can hold plenty of moisture.

With moisture and heating, we have two of the three ingredients for thunderstorms in the area east of the cold front and south of the warm front. There's lifting north of the warm front, too. The development of storms would depend on addition of the third factor, instability.

Summertime thunderstorms that are not associated with a front offer a simple illustration of how instability develops from the heating of air near the surface.

The morning weather is clear, but as the sun heats the surface, some lifting develops. Puffy cumulus clouds form. As the day wears on and the surface temperature increases, the cumulus grow larger. By afternoon some thunderstorms might develop. Both the instability and the lifting are provided by the heating of the surface. Lifting is from a thermal effect; instability develops as the low-level temperature increases during the day while the temperature aloft remains the same. Thunderstorms that develop in such a situation are usually isolated, because there is neither a generous source of moisture to make a lot of storms nor strong instability to higher altitudes. There's just not much circulation aloft in the summertime to move colder air over warmer air.

Widespread outbreaks of thunderstorms require more instability than is generated by heating on a summer day. Here we have to consider what is happening aloft in addition to the surface weather patterns.

The 500-millibar chart gives a picture of the circula-

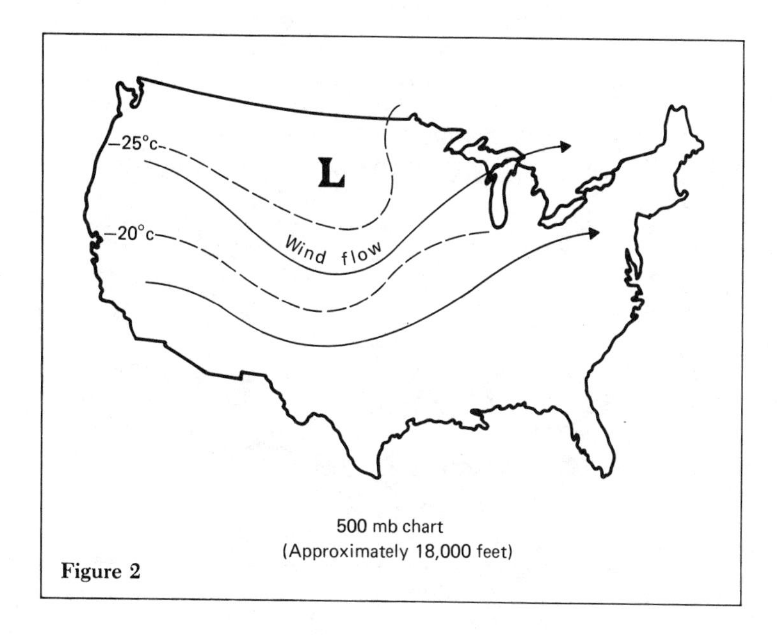

Figure 2

tion at about 18,000 feet. There can be lows and highs at this level, or undulating flows or straight east–west flows across the country. These patterns have a strong influence on our weather.

Instability is generated when colder air aloft moves over warmer air at the surface. If the patterns and characteristics of air at 18,000 feet were always the same as at the surface, this wouldn't happen. But the patterns are not necessarily the same, and more important, the temperature of the air aloft is not modified by the surface temperature. If the circulation aloft is from the northwest and around the south side of a low aloft, as in Figure 2, the air being drawn around the south side

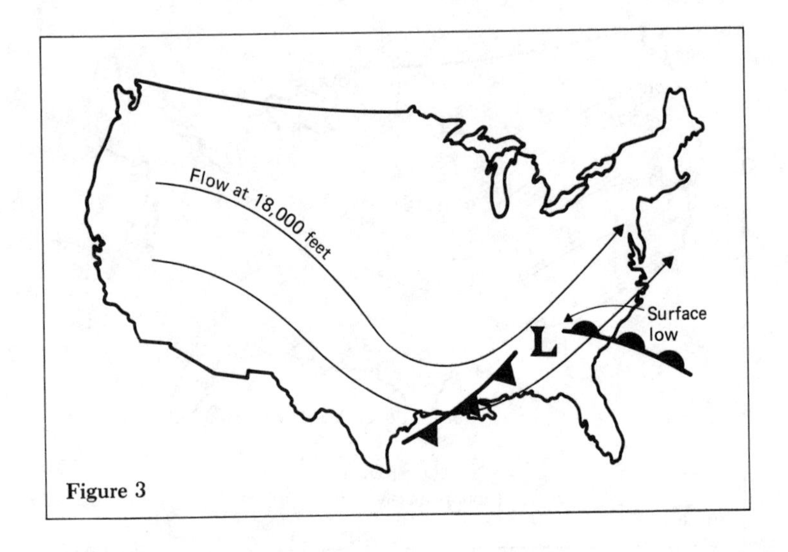

Figure 3

of the low will remain cold. Down below, the air ahead of a surface low might be warm, because it is drawn from a (relatively) warm area. This is a classic generator of instability. Figure 3 shows a 500-millibar chart imposed over a surface chart on a day when instability was strong and thunderstorms were numerous over the southeast.

A key to the study of thunderstorms is the advection of cold air aloft over warmer air at the surface. It was simply put by a very wise meteorologist: when there is a strong southwesterly flow aloft, consider the possibility of thunderstorms. There is a low aloft to the west, and cool air is moved down around south of the low and back up over warmer air to the east.

TAKE A LOOK

The textbook descriptions of how nature combines all the factors to make a thunderstorm are interesting. Before moving to that, though, put a thunderstorm in perspective by thinking back to the last time you flew by a building cumulus or cumulonimbus. Just looking at the billowing cloud carries the message of dynamic development, moisture, and turbulence. Looking straight down at the ground from the airplane emphasizes the energy released by instability. The higher you fly, the more impressive it becomes. I was recently flying at 21,000 feet around a monster storm, with about as much of it above my flight level as below. The ground was far away; so was the storm top. Some blast. Some energy. If we are going to share the air with stuff like that, we had *better* understand something about it.

BASICS

As air rises, it expands and cools. When it cools to the dewpoint (the level to which the temperature has to drop for the air to become saturated) condensation occurs and clouds form—if there is moisture in the air. This is why temperature and dewpoint are important to pilots. If they are together at the surface, clouds form there. Fog. If the temperature and dewpoint are close together, any clouds that form will be at a low level. If the air above the level of condensation is stable, stratus clouds will form. Stratus clouds don't develop into thunderstorms. But if the air above the condensation level is unstable—that is, if it cools at a rate of 2°C or more per 1,000 feet—then the clouds will continue to develop vertically. The condensation process gives off some heat, so the moist air in the cloud will cool more slowly with altitude than the surrounding air. When a particle of air is warmer than that surrounding it, the tendency is for it to continue rising, even to accelerate. The rising parcel of air has become unstable relative to its surroundings. Convection, building, will continue without any outside lifting influence. The lifting action got it started; from here it can grow on its own. This is called the level of free convection.

Instability does not have to exist all the way up for thunderstorms to develop. In fact, the temperature seldom decreases at a uniform rate. Whenever you are climbing in your airplane, watch the outside air temperature gauge and record the reading every few thousand feet to get a feel for the vertical composition of the atmosphere. Also, compare actual temperatures aloft with those forecast. If the temperature aloft forecast (included with the winds aloft forecast) is wrong, there's likely to be a lot wrong with all forecasts.

Because condensation releases heat, the moisture supply available to the building cloud is important. The more moisture drawn in, the more heat released at condensation and the greater the difference between the temperature of the building cloud and the surrounding air. This results in an acceleration of the upward-moving air inside the cloud, much as a longer shot from the burner of a hot-air balloon makes it climb faster. And the upward acceleration gives even more strength to the feed of moisture drawn into the cloud from lower levels. It builds on itself, rather like a slow-motion explosion. As you fly through, however, the motion might not seem so slow.

The inflow packs a lot of moisture into the cloud. It is only humid air until condensation occurs, then it becomes visible in the form of cloud. As the cumulus churns and grows upward, the moisture particles collide and unite into ever-larger water droplets.

As the cumulus cloud comes closer to reaching the mature cumulonimbus stage, rain might be encountered in the cloud even though none is falling beneath. It is at this time that the first radar return of the cloud might show. And shortly, the radar return will grow. The updraft won't be able to continue supporting the increased weight of the water droplets, and they will continue merging into larger drops. Rain will start to fall. The rainfall is likely to start from about the middle of the cloud—at about the freezing level—and it is not until lightning develops and the first rain falls out of the bottom of the cloud that the storm is considered to have matured. The level from which the rain first starts to fall is where the downdraft starts; it later spreads upward and outward to more of the cloud. Heavy rain has been encountered in a freshly matured thunderstorm as much as 10,000 feet above the general freezing level in the area.

As the storm matures, the updraft continues and might actually accelerate around the edges and upper part of the cloud. What appears to be one thunderstorm might actually be several cells in various stages of development.

Go now from this theoretical explanation of the birth of a thunderstorm back to the left seat of the airplane.

When flying around cumulus, the growth of the clouds is easy to see: they look like rapidly expanding cauliflower. The textbook description mentions only updraft in the first stage of development. This suggests that you'd get upward action and not much turbulence, but this isn't the way it works in practice. Any cumulus that is more than 5,000 feet tall contains plenty of turbulence as well as the updraft. Look at the cloud: you can actually see the turbulence. The cloud churns: for a fact, there is an updraft in there. But there is no updraft outside the cloud—actually there might be some subsidence (settling of air) in the surrounding clear air as the cumulus pulls low-level air in, so there is vertical windshear and turbulence around the edges of any building cumulus. This vertical windshear will increase as the updraft inside the cloud increases. Any time you have air moving in opposite directions, or at different velocities, there will be turbulence at boundaries with air moving differently or with air not moving at all.

The concentration and enlargement of water droplets inside is apparent when we fly through a cumulus, too. Even when the cloud is up to only 12,000 or 14,000 feet, there can be enough water to appear as light rain on the windshield even though nothing is falling from the bottom of the cloud. And by this time the turbulence is pronounced enough to make passage uncomfortable in a light airplane. Even big airplanes get bounced around; airline pilots are frequently heard calling with word that they are deviating around a "little ol' buildup." That only means they want to keep the coffee in the cups and the martinis in the pitcher.

When a cloud has reached 20,000 feet, it has a real billowy appearance. If we fly by at a level near the top, vertical development can appear as a rapid upward spurt. Awesome. Equally impressive is the fact that the cloud is made out of water, which isn't the lightest thing in the world. It takes a lot of energy to stack water that high, and when it starts to fall, you know something exciting will happen. It's even exciting before water starts to fall. It goes without saying that, whenever possible, wise pilots steer clear of cumulus clouds that have reached this level whether or not they are producing a radar return or lightning.

MATURITY

The turbulence in a cumulus cloud was caused by an interaction between an updraft and air that was either not moving vertically or that was settling gradually. Turbulence in a mature thunderstorm is a result of the interaction between air flowing into the storm and the resulting updraft, and air flowing down and out of a storm. This raises the ante far beyond what it was for the cumulus. For an example of what this might mean, consider that the updraft in the growing cumulus might have been working at an average rate of 15 feet per second, for a net difference of 15 fps or a little more between it and the clear air a few hundred yards away.

At the mature stage of the storm, the updraft might reach 60 fps or more and the downdraft 25 fps, for a difference of 85 fps. That very unscientific (but logical) example suggests that the turbulence passing through a mature cell could be almost six times as bad as the turbulence encountered when passing through a 14,000-foot-tall cumulus: any pilot who has nibbled at such a cumulus isn't too enthusiastic about multiplying by six.

A thunderstorm isn't really a thunderstorm until lightning begins. This generally occurs when the cumulus builds to a level where the temperature is about −28°C. That's pretty high; on a day with "standard" temperature it's 22,000 feet. Thunderstorms usually occur with above-standard temperatures, though, so lightning is more likely to commence as the cloud builds toward or through 30,000 feet.

DOWNDRAFTS

The downdraft comes with the first rainfall and increases in strength as the rainfall increases. It brings cool air down, which fans out over the surface pushing underneath the warm air that is still rushing into the storm. This causes lifting at the lower levels and contributes to the strengthening of the storm or the development of another nearby. The horizontal windshear in this area of interaction between the downdraft and updraft is very strong; at low levels it is characterized by the roll cloud that often forms ahead of a thunderstorm. The roll cloud's name is descriptive. As the air rushing into the storm reaches the condensation level, a cloud forms; the action between the outflowing and inflowing air causes the rolling. The turbulence would match the appearance of the cloud.

An individual thunderstorm cell has a finite life. From maturity it might go for an hour or two hours before the downdraft totally dominates the cloud and the storm is considered to be dissipating. But if the ingredients that spawned the first cell still exist, there can be continuous development, and a cluster or line of storms can last for a long time. Continuing activity is usually associated with frontal activity, but certain characteristics of storms can cause regeneration within a given air mass, with no frontal influence.

NEW CELLS

When mature thunderstorms are close to one another, the colder outflows displacing warm, moist air at the surface can provide the necessary lifting to trigger the development of a new cell. This new development would most likely come in the area between cells, or in the direction toward which the downdraft is the strongest (usually to the east or southeast of a thunderstorm in the northern hemisphere) or the direction in which the storm is moving (determined by the middle-level winds), usually to the east or northeast.

The development of cells between other cells is of special importance to pilots. This happens most frequently when existing cells are 3 miles or less apart, and can rather quickly convert what appears to be a narrow pathway into a maelstrom. There's a rule of thumb on avoidance of garden variety cells. It calls for a 5-mile berth, suggesting that the minimum gap between storms for a safe passage would be 10 miles. This guideline addresses both the question of turbulence around the outside of a thunderstorm and the possibility of new cell development between other cells. There's another guideline for avoidance of severe storms: 20 miles. This distance should be maintained for *all* thunderstorms when operating at altitudes above 18,000 feet.

BACK TO THE LEFT SEAT

It is often possible to observe the generation of new cells from the airplane. The back (west side) of an area of storms might consist of gray, amorphous shapes; the appearance of the east can be one of dynamic development.

I saw a classic example of this one day in the St. Louis area. I was flying high, at 21,000 feet, eastbound. Thunderstorms ahead seemed to be in clusters, with one short line trailing off to the southwest. Viewed from the west, the situation didn't appear ominous. Some truly billowing cumulonimbus were visible, but there were lighter gray areas.

At first I thought I'd fly through the area, deviating around storms with the help of airborne and traffic control radar. Nearer, this didn't look like a good idea. The radar returns were strong and suggested only narrow pathways between the storm cells. The air traffic controller said that most airplanes were detouring to the south. As a clincher, I heard a pilot who was trying to penetrate rather than avoid the area announce, in a tense voice, that he had to do a 180-degree turn and get out of there. It had apparently looked okay to him from a distance only to turn mean when viewed closer.

The detour was about 50 miles to the south and the edge of the activity. Once there, I could turn back toward the northeast. And as had happened many times before, the view of the storm system from the east was a reminder that they are usually meaner on the side toward which they move. There was a lot of new cell generation, and although from the west the system showed occasional signs of dissipating, from the east it looked robust, healthy, and growing.

TOPS

The taller the thunderstorm, the meaner the thunderstorm. When pilots hear of one with tops to 60,000 feet, they do indeed become interested in the location of the monster. As well they should.

The relationship of height to intensity is logical. When a thunderstorm builds above 30,000 or 35,000 feet, it is a sign of strong instability in the middle levels encouraging the push on up to abnormal heights. The stronger the updraft, the more moisture pulled into the cloud, the higher the tops at maturity, the more development after maturity, and the harder the rain and stronger the downdraft at maturity. Don't read this as suggesting that short storms are okay. I've heard pilots say that thunderstorms building only to 25,000 or 35,000 feet are relative runts and can be managed. This is a foolhardy approach. Any mature cell can be extremely turbulent.

Height usually relates to hail. It takes a lot of up action to get supercooled moisture up well above the freezing level and keep it there while droplets keep colliding and building an ever-bigger hunk of ice. The force of the updraft is evident, too, when you consider that the enlarging hailstone has to be held aloft until it reaches a size and weight that overpowers the updraft and lets it fall.

The fact that hail falls away from the rainfall has been used to suggest that an updraft is strong enough to throw hail out of a storm and have it fall as much as 5 miles downwind (as related to the upper level winds) from a storm. And while airplanes have certainly been battered by hail away from precipitation echoes, even clear of cloud, some meteorologists belittle the hail-throwing theory. Hail, it is said, falls where there is no rain because the large hailstones are heavier than rain and fall in different patterns. Also, the updraft shaft can be tilted because of wind, so hail could conceivably fall into clear air beneath the tilt of a storm. There's enough difference of opinion to get good arguments started among meteorologists on any number of thunderstorm subjects. Hail is one of them. As pilots, our best bet is to assume that they are all correct when identifying hazards. That will make us always err on the conservative side.

It is certain that there is probably hail in most all storms. Some might melt and fall with the rain of a storm; in fact most hail melts before it gets to the ground. The worst levels for hail inside the storm are between 10,000 and 30,000 feet.

BOTTOMS

The height of the bottom of a storm also has a bearing on flight operations.

On an average summer day the little cumulus cloud that grows into a thunderstorm probably starts off based at 3,000- to 4,000 feet. This base will remain relatively unchanged as the cumulus grows. Fly 400 or 500 feet beneath the cloud's base and you will certainly feel the updraft. The bottom of the cloud will often appear darker as the cumulus grows—a sign of greater moisture concentration.

Once the storm matures and rain starts falling, the base of the storm might seem to remain at 3,000 or 4,000 feet for a while. But as the rainfall rate increases, the effective ceiling probably decreases. The cold air, the downdraft, and the rainfall change both the temperature pattern and the moisture content of the air beneath the cloud. The 3,000- to 4,000-foot base from which the cumulus is building might not seem to change around some edges of the storm, but underneath a mature storm the ceiling can be indefinite in heavy rain, and cloud might form in there as well. Also, the bottom of the roll cloud might be below the base at which the original cumulus cloud formed, because the cold air of the downdraft lowers the condensation level. This cooling is what prompts the formation of other low-level clouds beneath and around a thunderstorm. Bases of thunderstorms can also be affected markedly by frontal slopes, which we'll explore a bit later on (pages 24–29).

TURBULENCE

Taking the basics reviewed thus far, consider where the greatest turbulence will be found around a cumulus as it grows into a cumulonimbus. Certainly there will be a lot of turbulence in the storm early on. There is an interaction between the downdraft in the center, the updraft around the downdraft, and the clear air outside the cloud that's moving neither up nor down at an appreciable rate. This is shear turbulence, caused by adjacent air doing different things. Above levels from 3,000 to 6,000 feet (in garden variety storms) this is more likely to be vertical shear—that is, up, down, and relatively still air rubbing against one another and causing substantial turbulence.

At lower levels, the downdraft fans out because it sure can't blow through the earth; the shear thus starts to become more horizontal. Cold downrushing air fans out beneath warm and humid air that is being drawn into the storm.

Flying into the side of a storm up high, the sequence might be no draft, updraft, downdraft with strong turbulence before and as the updraft is penetrated, and very strong turbulence in the area between the updraft and downdraft. Down low the airplane might go from flying with a strong tailwind (and some updraft) into an area with a strong headwind (and some downdraft) if approaching a storm from the direction toward which it is moving. In any case, the pilot would experience turbulence as well as fluctuations in airspeed and probably altitude.

How much turbulence might be expected in and

around thunderstorms? There is no set answer to that question other than "a lot." The nature and amount of turbulence depends on many things, and can change rapidly as a storm goes through development, maturity, and dissipation.

The strength of the updraft has as much to do with turbulence as anything else, because stronger updrafts make taller clouds and taller clouds make stronger downdrafts for a greater total difference between the up and down action.

Updrafts are not measured in storms (except in research flying, which will be discussed later), but in a way downdrafts are, in the form of surface observations of the peak gust as a storm passes a reporting station. It's not uncommon for gusts to reach 50 knots in a passing severe storm. That's about 84 feet per second, or about 5,000 feet per minute. (The actual downdraft within the storm would be somewhat less than that, because the strength of the first gust is enhanced by the storm's motion across the surface.) When you imagine such a storm downdraft and then consider that the updraft is an even stronger flow and that they are in close proximity, you will have some idea why thunderstorms are dear to every pilot's heart.

The pilot's primary concern is usually with turbulence, not with precipitation. And the radar pilots (or traffic controllers) have does not show turbulence. Nor does a Stormscope (a device that displays the location of electrical discharges). Radar shows only precipitation, and the actual area of rain might not be where the turbulence is at its worst. The heaviest rain is associated with the downdraft, where it might be relatively smooth. What any form of weather avoidance gear *will* show is where—or in what direction—a thunderstorm is likely to be. The chore becomes staying a proper distance away from storms in order to avoid the chaotic conditions in and around them.

Remember, the greatest probability of severe turbulence is around the edges of the storm, where there is an interaction between the surrounding air and the updrafts and downdrafts created by the storm.

FRONTS AND THUNDERSTORMS

The classic picture of frontal thunderstorms involves a squall line preceding a cold front. There's a flaw in this picture, though. The really fancy squall lines are not found right along cold fronts, where colder air is pushing in under warm, moist air at the surface. Instead, the big ones are usually found 100 or more miles ahead of a cold front. This is the result of a circulation within the jet stream, an area of very strong wind aloft, high above the surface position of the cold front.

FRONT WITH JET STREAM

There is circulation around the core (where the wind aloft is at its highest) of a jet stream that results in descending air in the right front quadrant of the area. This descending air results in heating and pushes unusually warm air out ahead of the cold front. This push can result in convergence ahead of the surface position of the cold front; this combines with basic instability in the warm-sector air to get upward motion started. There is generally colder air aloft to the southeast of the jet stream, and this provides the additional impetus of instability for the development of a squall line containing severe thunderstorms. And the situation persists for as long as the conditions remain. There is a continuous generation of cells along the line. Such a squall line might be called frontal, but the relationship is more with the jet stream than with the surface front. The front is there. It is part of the pattern, but it is not in itself the trigger for the squall line, which doesn't develop until the jet stream is in the right place.

An interesting proof of this comes when you examine the 500-millibar charts for an active thunderstorm time in the spring. By identifying the days on which there is a very strong southwesterly flow aloft, you can do a pretty good job of locating squall lines without even looking at the surface weather charts. On the other hand, when looking at the surface charts without knowledge of upper-level patterns, it is apparent that there are a lot of cold fronts without attendant squall lines. Figure 4 shows two examples of this.

The relationship between the jet stream, thunderstorms, and thunderstorm forecasts was clear on a flight from Denver to Trenton.

A cold front was located in central Illinois. Wind forecasts called for very strong northwesterlies at 18,000 feet over Kansas, swinging around to very strong southwesterlies over Illinois. The forecast was 240 degrees at 84 knots at 24,000 feet over Illinois. Widespread thunderstorm activity was also forecast. Good news and bad news. The tailwind would make a long trip a lot

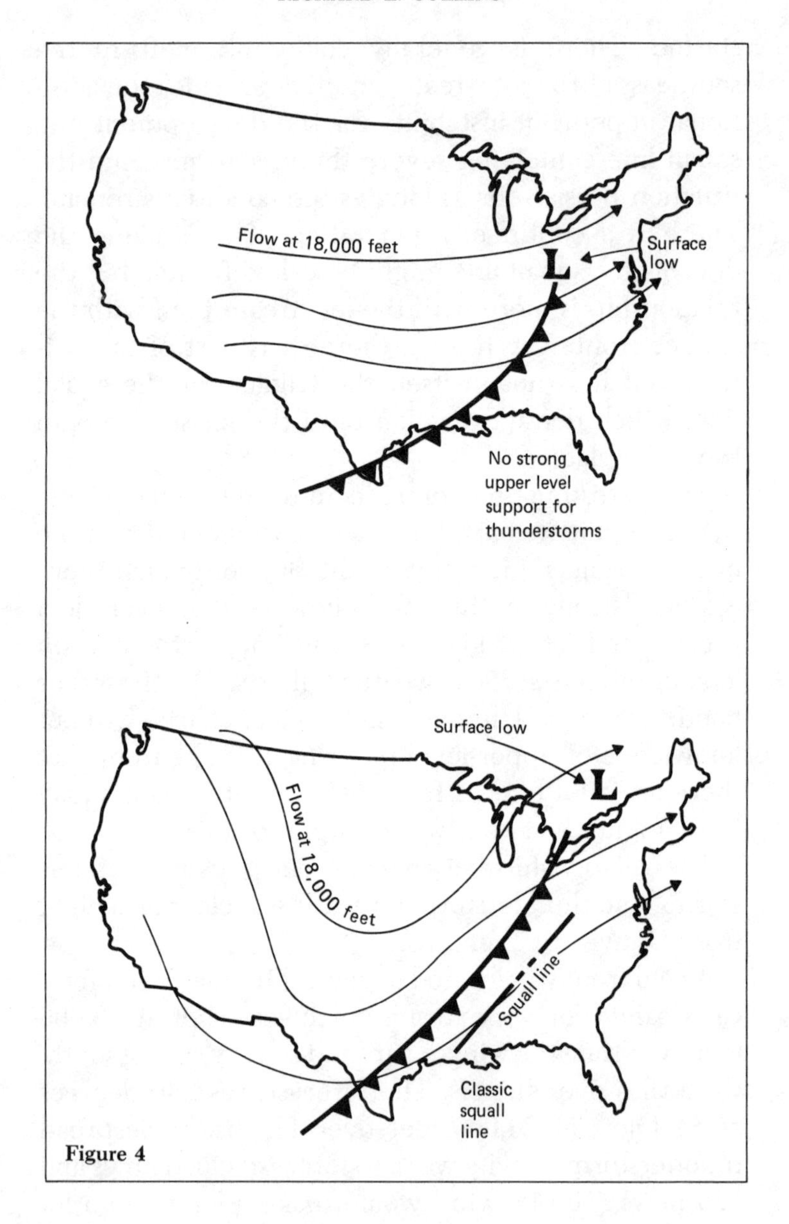
Flow at 18,000 feet
Surface low
L
No strong upper level support for thunderstorms
Surface low
L
Flow at 18,000 feet
Squall line
Classic squall line

Figure 4

shorter, but the pattern was one that could breed a squall line. It, in turn, could dictate a long detour.

The first leg was from Denver to Moline, Illinois. The weather was breathtakingly clear leaving the Denver area and it remained that way to Moline. The tailwind, however, wasn't all that good. It started off as very light but did increase some right before a descent into Moline was started. On that descent, a long gray line aloft was visible to the east. It looked for all the world like a squall line.

The surface wind at Moline was southwesterly, indicating that the cold front had yet to pass. It was a classic setup for a squall line ahead of a front.

The weather briefer said there was indeed some activity to the east but that the heavier stuff appeared to be south of our course.

Back aloft, at 21,000 feet, the tailwind was achieving good proportions—over 70 knots—as we approached that long gray line sighted earlier. There was no weather being painted in the area, though, and on close visual examination the "line" proved to be other than cumulonimbus. There was a cumulus appearance, with some of the clouds having that fearsome mammatocumulus look, but the clouds were based at 17,000 or 18,000 feet with tops up around 30,000. We headed for the area with the most amorphous look (as opposed to the tough-looking mammatocu) and had a smooth ride through. Immediately on entering the area, the excellent tailwind faded. What jet stream activity that existed had been of short duration, and apparently the upper level support for a real line of thunderstorms simply did not exist in that area. There was extensive thunderstorm activity ahead of the front to the south, where moisture supply was apparently more plentiful, instability more pronounced, and southwesterly winds aloft somewhat stronger. Up our way, there was a more widespread band of rain and rainshowers, spawned by a gentler set of circumstances than had been anticipated.

FRONT WITHOUT JET STREAM

Even when a cold front doesn't get the strong support of upper-level circulation for storm generation, there can still be thunderstorms. And they can be strong enough to deal misery to a pilot penetrating the area. The lifting mechanism of the cold front itself can combine with moisture and instability to generate strong, if not severe, thunderstorms. Every front should be considered as a potential source of thunderstorm problems, and each should be judged on its merits.

An observation that is usually valid relates to the relationship between the width of the area of cold-frontal inclement weather and the severity of that weather. Where there is a line of strong thunderstorms ahead of a cold front, the line is probably only 10 to 20 miles through. That basically means that all the energy is being expended over a relatively short distance; this can be an indication of violence aloft. By the same token, if the area of precipitation related to a cold front is 50 or 100 miles through, then the situation might not be so violent. It's rather like choosing between a sudden stop and a stop that consumes a bit more distance. The latter will probably be more comfortable and was what had actually been encountered on the flight out of Moline.

Another example of cold-frontal weather spread over a distance and lacking strong upper-level support was encountered on a late-spring flight across Texas. The flight was from Kerrville, Texas, to Dothan, Alabama, and as you can see in Figure 5, this route crossed a cold front at a rather oblique angle. The accompanying 500-millibar chart shows that there was no strong upper-level support for this system. The upper flow was mostly westerly across the area, and at 7:00 A.M. the temperature at the surface was +20°C; at 18,000 feet it was −13°C for an average drop of slightly less than 2 degrees per 1,000 feet. There was thus no strong in-

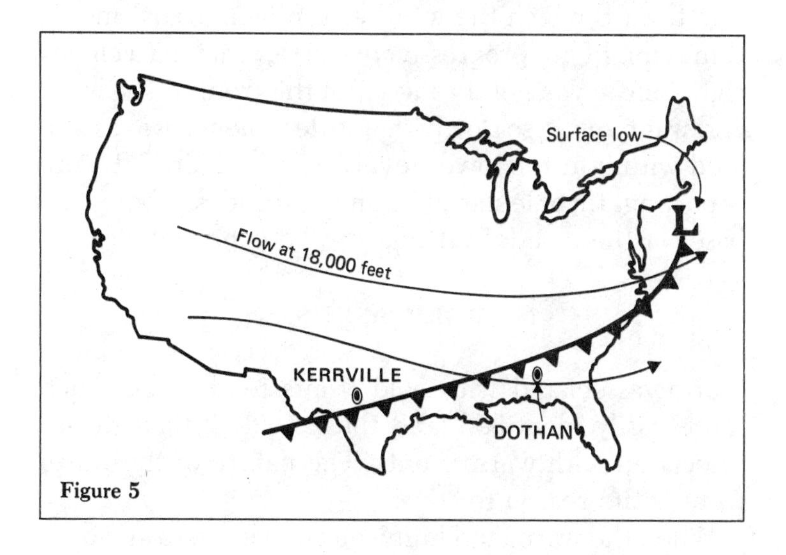

Figure 5

stability over the total range of altitude. But when you consider that the air was stable up to about 6,000 feet, instability above that level was logical. This was plain to see from the ground. Cloud bases were high, rain was heavy in spots toward the east, and there was some lightning. It was a pretty picture of the back side of a rather weak front.

The area of rain was about 100 miles through when penetrating the front at a 30-degree angle. Perpendicular to the front it would have been a little shorter trip through. The distance, the lack of upper-level support, and the high cloud base suggested that there wouldn't be any real bad spots in it down low, and there weren't. Of course, a pilot has to be wary flying from the back to the front of a thunderstorm area; the fireworks associated with the front are most likely to be found on the side toward which the weather is moving. That means things might get progressively worse, reaching a climax right before you get through. But this front was so weak and was moving so slowly that little violence was associated with it at the lower levels. Some jet aircraft were reporting turbulence at higher altitudes, above the bases of the clouds, but that's to be expected.

WARM FRONTS

Storms associated with cold fronts tend to be much more easily identified (and thus avoided) than storms associated with warm fronts. The nature of the warm front is the reason for this.

When the warm and humid air moves up to and over (heat does rise) cooler air to the north, the demarcation line at the surface is called a warm front. This usually has few of the clear-cut characteristics of a cold front, where cold air pushes under warm. The slope of a warm front is shallow, with an average of 1:100 or a little more. Textbooks put the possibility from 1:50 all the way up to 1:500, which seems more like a range to cover every eventuality rather than a real probability.

South of the surface position of the warm front and well east of the cold front, cumulus clouds are likely to be floating around, often in a reasonably blue sky. There might well be some scattered thundershowers. It's nice warm and humid weather. The surface wind is out of the south or southwest.

North of the surface position of the front, the surface wind is more easterly and the weather is strongly influenced by warm and humid air overrunning the cooler air near the surface. Winds aloft are usually out of the southwest above the slope of the front. As the warm air is lifted, condensation occurs. There's plenty of moisture to form clouds, and the area north of a warm front is usually cloudy and rainy, with low ceilings caused by warm rain falling into colder air near the surface. (The same principle as running a hot shower in a cold bathroom.)

With the slope of a warm front at 1:100, warmer air will be about 500 feet above the ground when 10 miles north of the surface position of the front and about 5,000 feet when 100 miles north of the front. That *is* a shallow slope, but not so shallow that there's not enough lifting to get some thunderstorms started. If the slope is 1:200, then the warmer air will be up to 5,000 when 200 miles north of the front—a very gradual lifting.

Warm-frontal thunderstorms can be embedded in other clouds and are often surrounded by an area of general rain. And while they are generally not as severe as cold-frontal storms, the fact that they are harder to see makes them a hazard, especially to light airplanes operating IFR.

VIEW FROM THE SOUTH

One key to warm-frontal activity is the stability and of moisture in the air south of the front. If that air is stable and relatively dry, as witness no thunderstorms and not a lot of cumulus clouds, then the possibility of warm-frontal storms isn't so strong. Neither is the possibility of a lot of warm-frontal weather. If there's little to build in, there might not even be a warm front drawn on the map out to the east of the low.

If there are scattered thunderstorms in the area south of the warm front, the potential for real trouble is much more serious to the north of the front. Once the warmer air gets the added impetus of lifting over cooler air at the surface, things can really get going.

VIEW FROM THE NORTH

Going far to the north for a moment, there's a way to judge the relative stability of overrunning air that is part of an approaching warm front. If the first high clouds signaling the approach of a warm front are flat, delicate cirrus, things may be relatively stable. But if they are cirrocumulus—high clouds with some vertical development—then the overrunning air is unstable. Cirrocumulus, incidentally, are characteristic of what sailors used to call a mackerel sky, a portender of bad weather.

OVERALL VIEW

So there are clues to the possibility of warm-frontal storms when viewing the situation from south of the front or well north of the front. Away from these two positions things become much less clear. There can be general rain over an area with storms embedded in the rain area. Or the storms can be hidden by other clouds. The location of storms in relation to the position of the surface front can vary, too.

Surface weather conditions north of a warm front are usually inclement to the point of precluding VFR flight, so pilots flying IFR do most of the worrying about warm-frontal storms. Knowing the surface position of the front and relating this to the average slope of a warm front is an important part of identifying problem areas and avoiding storms.

The cool air beneath the warm-frontal slope is stable. So there is nothing much going on between the ground and the slope of the warm front, where the warm air is overrunning the cold. Any thunderstorms that develop will be based on the slope of the warm front, and the colder and more stable air beneath will mitigate the incursion of the storm's effects (other than the rain) into the lower levels. The storm's feed is from above the frontal slope, and while the downdraft will probably penetrate into the air beneath the front to some extent, the shear turbulence caused by interaction between outflow and inflow close to the base of the storm shouldn't be pronounced in the lower levels.

Given a 1:100 frontal slope and unstable warm air beginning about 5,000 feet above the ground 100 miles to the north of the warm front, there is some room for low-level IFR in stable air. Two hundred miles north of a 1:100 front there should be 10,000 feet of pretty good air.

The trouble with all this is that the National Weather Service doesn't include slope information in forecasting warm fronts. In fact, the precise location of a warm front (as marked by an increase in temperature, a shift in wind from an easterly or southeasterly direction to southerly or southwesterly, a reversal of a dropping trend in atmospheric pressure, and a decrease in dewpoint representing drier air) is often difficult to pinpoint. Often there is no warm front depicted on a weather map to the east of a low-pressure area because there is no clearly identifiable position for the front. Rather, the condition is ragged, with inclement weather covering a large area. In this case, the slope of whatever front might try to identify itself is probably shallow. As with cold fronts, it is a rule of thumb that the larger the area of weather associated with a specific front, the less the likelihood of more severe weather. The exceptions to the rule, related to one's proximity to low-pressure systems or low-pressure waves, will be covered later.

I've never seen it written, but it is certainly logical that the higher the base of a warm-frontal storm, the less the intensity of the storm. It simply wouldn't grow as high before maturing and starting to dissipate. Conversely, the closer you get to the surface position of the front, the stronger the intensity of any storms that might develop.

The general rainfall north of a warm front falls from the clouds based on the frontal slope. The clouds that form below are a result of the rainfall (warm rain into cooler air) and are not generally rain generators themselves. But these lower clouds can obscure bad things ahead. Flying south toward a warm front at 5,000 feet, you would encounter a situation with a high overcast first. Rain would start two or three hundred miles (or perhaps even a little more) from the warm front. This first rain would be falling from clouds 10,000 to 15,000 feet high. There would be no turbulence. A bit farther on, lower clouds might be encountered. Flight on top of these and beneath those on the base of the front could be possible for a while. Rain would increase as southward flight continued. The upper clouds would begin to lower.

One clue to instability above the frontal slope might come from the nature of the rain. If it starts as a gentle and steady rain and slowly escalates to a harder but still steady rain, that could be a general sign of a more stable situation. Likewise, if the rain begins as showery in nature and becomes more intense as the airplane draws closer to the surface position of the front, that could be a sign of instability aloft and a signal to watch for thunderstorms.

I've heard pilots say they flew through a thunderstorm and there was nothing to it. There's likely a basic error in such an observation. The flight was probably beneath a warm-frontal thunderstorm. The heavy rain was there, as was the lightning, but there was no turbulence. It would be quite hazardous for a pilot to complete such a flight without a bobble and use the results to form an opinion of *all* thunderstorms.

COLD FRONTS

Cold fronts are easier to describe than warm fronts because they are more clear-cut. The cold front's slope is steeper. When compared with a warm front, it produces the same amount of lifting over a shorter distance. The active cold front is accompanied by a narrower band of clouds and precipitation and is likely to be more violent. One thing can have a very direct effect on the violence part, though. Proximity to a low-pressure center has a lot to do with the development of thunderstorms. And when operating north of a warm front, you might well be in close proximity to a low. The classic low generally moves in a direction parallel to the isobars in the warm sector (south of the warm front and east of the cold front), and a low can generate some pretty mean weather in the direction toward which it is moving. So you need to know the location of the low as well as of the fronts.

LOW BLOW

Circulation around a low is counterclockwise and inward. Inward in this case also means upward: if the upward part doesn't exist, then the low-pressure area will fill and cease to exist. If a low is maintaining its strength or deepening (as shown by a continuing reduction in pressure), then we know the upward part, the lifting, exists, and is stronger when closer to the low. This means, very simply, that if the other conditions conducive to thunderstorm formation exist—moisture and instability—the circulation into the low can provide lifting to trigger development. And when the other basics are added, it's clear that the area ahead of an advancing low is worthy of study. There's a southerly or southeasterly flow, so there is plenty of moisture. And the upper air patterns that usually exist when a strong surface low is working can bring cooler air in over warmer air at lower levels to generate instability.

This again has to be related to the jet stream. In discussing the formation of squall lines ahead of a cold front, descending air in the right front quadrant of the jet stream was cited as a trigger mechanism for the formation of squall lines. Well, every downer has an upper. There has to be some ascent of air to go with that descending air in the right front quadrant of a jet stream. The pertinent ascent in this case is found in the left front quadrant of an area of jet stream activity. And it is under this quadrant that surface low-pressure areas can strengthen and be sustained. Figure 6 shows a 500-millibar chart with a jet stream imposed over a surface chart on a day when a strong low developed and when there was both a squall line and the development of thunderstorms to the northeast of the low. Pilots usually worry only about the route they are flying, but the big picture is important as well: the development of a squall line many miles to the south, ahead of a cold

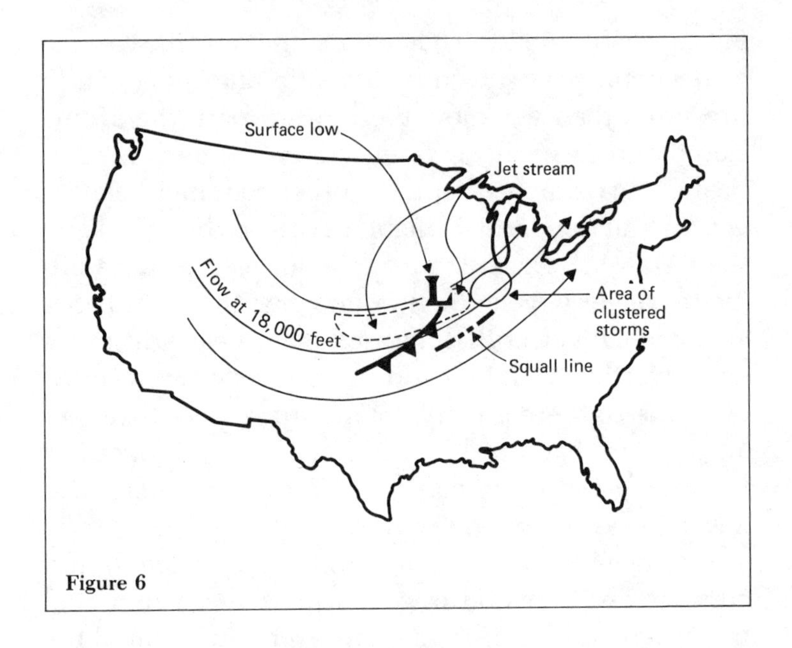

Figure 6

front, might give some signal about things to come from an approaching low-pressure center. If there is a squall line down south, there's likely to be some jet stream activity. The conditions that caused that might, in turn, help intensify a surface low and the weather in its path.

CLOSED LOW

Low-pressure areas aloft have other effects on the development of thunderstorm activity.

When the circulation around a low at the 500-millibar level becomes complete—that is, when the wind goes all the way around it in a counterclockwise direction—it is said to be a closed low. A closed low at the 500-millibar level isn't a frequent occurrence over the United States, but when one does develop it can bring a marked change to what we think of as normal weather patterns.

Closed lows aloft tend to move slowly, and to slow the motion of surface lows. A surface low can thrive on the support of that low aloft; as the air swirls around and upward in relation to the surface low, it finds a place to keep swirling and ascending in the upper-level low, which is usually found to the west of the surface low. As the surface low deepens, the circulation can be quite strong and the fronts can move rapidly around the low. Thunderstorms can be very much in evidence, especially in the day or two after the surface and upper-level lows develop and start interacting.

OCCLUDED FRONTS

A rapidly strengthening low moves rather slowly, but as circulation around it increases, so does frontal speed. This can result in the faster-moving and more dynamic cold front overtaking the warm front, forming what is called an occluded front. The time of occlusion, where the surface position of the cold front reaches the surface position of the warm front, is a time when things become a little like a cold front and a little like a warm front. Strong and numerous embedded thunderstorms might rule for a while, until the cold air close to the surface, both behind the cold front and ahead of the warm front, mixes up at lower altitudes, and until the warm air is lifted and spread out enough for the whole thing to lose its potential for turbulent weather. An occlusion where a cold front overtakes a warm one is working toward a stable situation, with warm air above cold air. It's as the process starts that the potential for really turbulent weather exists. Too, a low center associated with a fresh occlusion is likely to be a strong one, which suggests trouble. Once that low develops fully, its speed across the surface will increase, and such systems can develop into fast-moving areas of really tough weather.

STATIONARY FRONTS

When a front stops—that is, when neither cold nor warm air is advancing—it is logically called a stationary front. (The term static front has also been used, but that's a bad name because static implies thunderstorms, which isn't always the case.) If nothing is really moving, there shouldn't be any lifting action and thus no thunderstorms, right? Not quite, because a stationary front can be the trigger mechanism for low-pressure waves that move along the front and cause lifting and thus thunderstorms. Stationary fronts are not all benign.

It's relatively easy to tell when a cold front is going to stop. Just hold up a handkerchief, or determine the surface winds through more scientific means. When the wind behind a cold front isn't perpendicular to the front (that is, northwesterly in the case of a cold front with northeast–southwest orientation), then the front is starting to stop. If the surface wind moves around to the northeast, more or less parallel to the orientation of the front, then the front has probably all but ceased moving. The basics of circulation tell why: in Figure 7, the low is strong, with a good circulation. Both the low

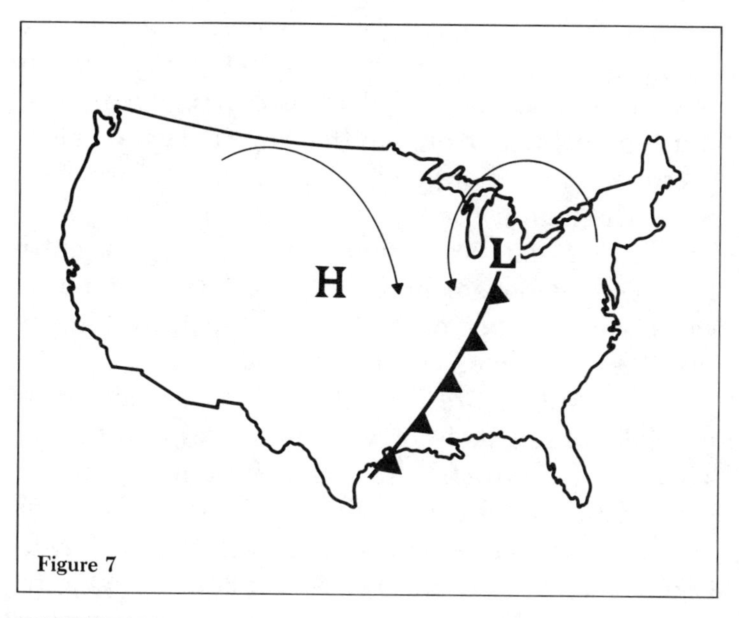

Figure 7

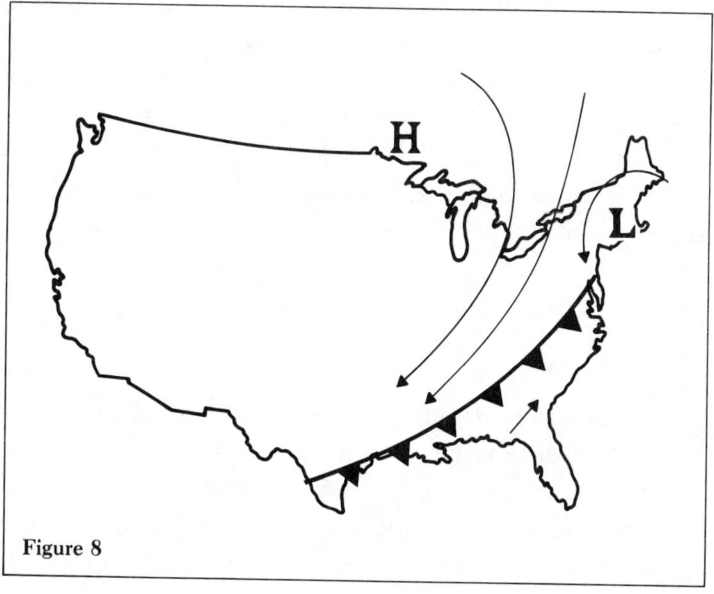

Figure 8

and the front are moving well, nudged along by that high-pressure area out to the west. In Figure 8, though, the situation has changed. The low has sped northeastward (probably tracking up the eastern flank of a low-pressure trough aloft) and is leaving a lot of front behind. The high has moved to the north of a lot of the cold front, and its circulation is thus no longer a contribution to the movement of the front. The flow is northeasterly on one side of the front and southwesterly on the other: a balance of forces.

There is still moisture. Clouds and rain remain. Some instability is still likely, or at least the prospect of some. This stationary-frontal situation often develops in the wintertime, when the upper air patterns go through periods of undulation across the country and the advection of cold air southward ebbs and flows. What a stationary-frontal situation waits for is a trough aloft to move cold air down over the relatively warm and moist air just to the east of the stationary front. This not only creates instability, it can help create a new surface low-pressure area that will change the circulation pattern and get things moving. If a strong jet stream develops, it can make the rebirth of that stationary front quite an event.

WAVES

In the lower levels, the northeast wind rubbing against the southwesterly can result in action. The ripple that develops in the stationary front is appropriately called a wave. It is trying to form a full wave, like a breaker at the shore. If it does, then a full low-pressure circula-

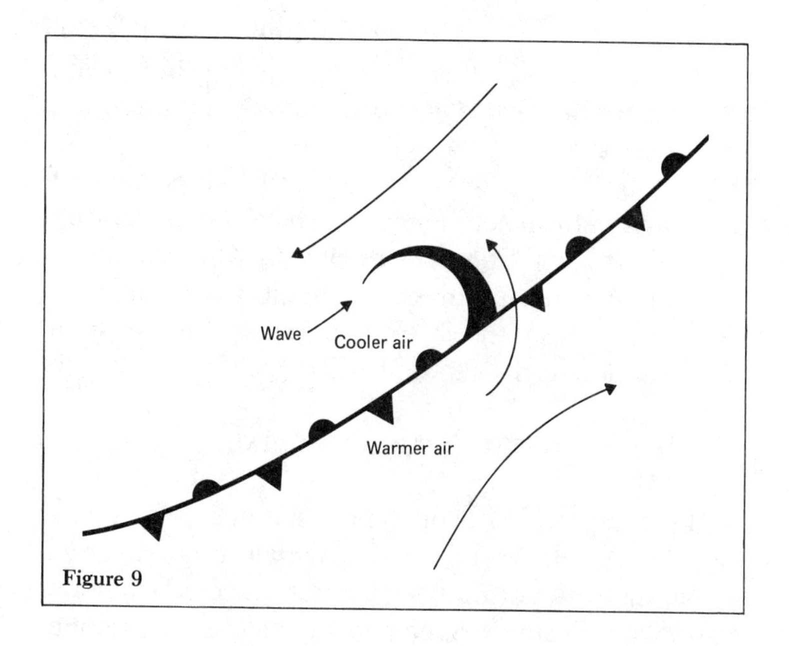

Figure 9

tion exists. If it doesn't, the wave will just move along the front, to the northeast, and finally dampen out, only to be replaced by another one.

As the wave forms, there is mixing of warm air and cold air, as in Figure 9. Heat rises, so the warm air from the southeast is forced up over the cooler air to the northwest of the front. Presto, lifting. If there is instability, then the lifting might well accelerate and thunderstorm activity might begin even though the frontal system itself is remaining stationary.

Low-pressure waves can form on stationary fronts anywhere, but perhaps the most frequent and classic effects of this phenomenon are found in the south central and southeastern United States, in late winter and early spring. There's plenty of moisture from the Gulf of Mexico on the southeast side, and by this time of year the cold fronts start to lose their punch and often stop in this area.

Such stationary fronts can remain for days, generating thunderstorm activity over a large area on a rather continuous basis. They are usually the direct cause of those continuing and massive rainfalls that come occasionally in the southeast, flooding everything in sight. The thunderstorms can be severe.

GETTING THINGS MOVING

Nothing stays still forever. What a stationary front waits for is the development of a low, with a complete circulation, that can get things back into the cold front and warm front business. Such a low develops on a stationary front when it gets the necessary upper-level support. The southwesterly flow remains steady while the front is stationary. When some jet stream activity comes round the bend of the trough aloft, it will give support for the development of a surface low. Remember, there is ascending air in the front left quadrant of the jet stream. That can be enough to let a low-pressure wave develop into a complete circulation and get things moving again. Indeed, a likely location for the development of a "new" low-pressure center is near the tip of a trough aloft. When such a low develops and things get moving, conditions often lead to a squall line ahead of the cold front as well as the development of thunderstorms north of the warm front and ahead of the low.

Regardless of the situation—cold front, warm front, whatever—the key to understanding thunderstorm development is in going beyond surface observations, to get a feel for the vertical composition of the atmosphere. With the southeast or east wind north of a warm front, visualize the overrunning southwesterly flow and the possible instability in that overrunning air. South of the warm front and east of the cold front, the jet stream holds the key as it moves cold air over warm and as the circulation within the jet stream works to spawn thunderstorms ahead of the cold front. After a cold front passes, stability is created by colder air moving under warmer air. When a front is stationary not far south of your position, the surface wind might well be northeasterly, but aloft the southwest wind might still hold sway. Not much here for thunderstorm development unless waves form on the front to provide lifting, or a complete low circulation develops and reactivates the frontal activity. In the path of a low, the airflow is moving into the low and upward. If the situation aloft is stable, with no trough or low aloft to support the one at the surface, it will fill, disappear, and no longer provide the lifting part of the thunderstorm equation. On a hot summer day the drop in temperature with altitude gives a clue to the possibility of thunderstorm development.

The situation aloft is also a key to the severity of thunderstorms and is the basis for severe thunderstorm and tornado watches and warnings.

RICHARD L. COLLINS

TORNADOES

Tornadoes are associated with fast-moving, dynamic weather systems. Springtime is the prime season. Warmer air is starting its push northward while the upper air patterns are still moving cold air southward. Tornadoes are always associated with thunderstorms and, naturally, are most common where thunderstorms are strongest.

People who live in tornado country develop a feel for the storms. Something about the still and humid air and the color of the sky gives a tip and, on the ground or in the air, alertness is a key.

My family and I lived in Arkansas for a number of years. In that time, and in years of flying in the Middle West, I saw a lot of examples of the dynamic nature of tornado-producing thunderstorms.

A tornado warning was posted one afternoon when I was working at home. While the sky looked friendly enough, I kept a good watch out the window. The vigil was rewarded late in the afternoon when dark clouds appeared on the horizon. I went outside for a look.

There appeared to be two thunderstorm cells, one to the west and one to the southwest. The bases of the clouds appeared rather high—probably 3,000 to 4,000 feet—and the area between the two clouds looked rain-free, though dark. An almost constant lightning was pulsating in the clouds. Our house was on the west side of a hill, and for a better look over the bushes around the yard I hopped atop the picnic table.

The storms were clearly moving closer, and in a minute or so I saw what I thought was a flock of birds, circling. A closer look suggested differently, unless new breeds of square and rectanglar birds were on the wing. It was clearly debris—shingles and doors and things like that. I went in and opened all the doors in the house (as is recommended in tornado situations, to allow pressure to equalize) and went back out to see if what I had seen was really a tornado, and if it was tracking toward our house.

Fortunately for us, the storm was tracking to the south. I stood and watched it go by (the nearest damage was several hundred yards south) and then waited to see how long it would take the sirens to start. It wasn't long; the twister had left quite a trail. The buzz of chain saws persisted for days as people cleaned up downed trees.

The size and character of this storm strongly supported the frequent admonitions to give severe thunderstorms a 20-mile berth. Even though there was a rain-free area between two cells that suggested the possibility of a visual passage, and which would probably have appeared clear on radar, the turbulence in that area would have been extreme. And the very nature of the strong feed and downdraft, coupled with the development of the tornado, could have made the whole area chaotic. I later looked at a photograph of the weather radarscope made during the height of the storm. There were two hook echoes (indicative of a tornado); one touched the ground and the other apparently remained aloft. They were on the south side of the storm.

RICHARD L. COLLINS

FAST

Tornado-spawning thunderstorms can develop very rapidly and in less suspect areas, as I learned one day when out photographing storms. The forecasts for this day called for storm development about 100 miles to the northwest of Little Rock, so we flew up that way. The area was pretty disappointing from a photographic standpoint. There were few storms around, and they were beginning to dissipate. So we started home.

On the trip up, there had been an area of building cumulus just to the northwest of Little Rock. As we returned, the air traffic controller said that some weather return was starting to show a few miles northwest of Little Rock, and as we got closer it was plainly visible. The clouds had that black, churning look, and I opted to give the precipitation good berth.

The detour turned out to be extensive. Cells were beginning to develop in a line and were quickly becoming intense in appearance. We were taking pictures all the while and got some of the beginning of a twister that ripped a path through Cabot, Arkansas, about 45 minutes later. It was a clear illustration of an explosive weather situation that developed 100 miles from where it was expected—the error probably caused by a little shift in the path of the jet stream.

One final example. I left Wichita bound for Houston at about ten one morning, with a forecast of pretty good weather. A light-to-moderate southwesterly wind was forecast at 18,000 feet. There was little showing on radar at the time. The latest available weather map positioned a surface low to the west and a warm front to the south.

En route, the weather was okay, with only a line of rapidly building cumulus between Oklahoma City and Dallas. But the wind aloft was very much stronger than forecast. At one point we had a 60-knot headwind component plus a lot of drift to the left. And we had some clear-air turbulence, especially when picking through the line of building cumulus, but the trip was routine except for the strong wind aloft. Things were cranking up for something big, though; a few hours later a monster tornado flattened Wichita Falls, Texas. When I heard about the tornado, I remembered the rule of thumb about warm air at the surface and a very strong southwesterly flow aloft meaning trouble. The actual upper winds are measured twice a day, and forecasts are based on likely developments. The forecasts were off this day; as a result, the warning of possible severe weather came later rather than sooner.

SPECIAL

Tornadoes might seem easy to anticipate because they are a product of excessive instability, which is a natural product of cold air over very warm and humid air. This situation usually produces only thunderstorms, though. It takes a special set of ingredients to generate enough instability to get a tornado going. Unseasonably warm air at the surface, an adequate moisture supply, and a strong advection of cold air aloft over the area are all required. This all usually comes together about 150 miles ahead of a surface cold front, usually in the spring. Once it gets going, the tornado itself is within a "tornado cyclone," a small but very strong low-pressure area that is perhaps 10 miles in diameter. Everything about it and near it is severe, in spades. And where pilots might feel okay flying relatively close to garden variety thunderstorms, there can be some nasty surprises close to a severe thunderstorm. The area affected is much larger because stronger instability results in a stronger feed into the storm, and the development of the cyclonic system that accompanies any tornadic activity that might develop would enlarge the area of severe shear turbulence all thc way around the storm. The rule is to stay 20 miles away from any thunderstorm cells when severe activity is forecast. Perhaps that admonition errs on the conservative side. But perhaps it does not in some cases. The lesson would be a hard one for trespassers.

When the National Weather Service identifies the possible development of strong instability coupled with the other necessary ingredients for thunderstorm formation, it issues a severe-thunderstorm or tornado watch. Such is advisory in nature—it tells a pilot that any storms that develop have the potential of being severe. Sometimes nothing happens, but the watches should never be taken lightly. When storms do develop, the NWS upgrades the watch to a severe-thunderstorm or tornado warning. Locations are pinpointed and the message is clear.

AIR MASS THUNDERSTORMS

When there's no identified front or low-pressure system in or near the area, any storms that develop are referred to as air mass thunderstorms. This simply means they form where there's no force to provide lifting other than what develops within the given air mass. Summertime storms are often of this variety, as are nocturnal storms.

SUMMER STORMS

An occasional isolated summer storm is probably caused by nothing more than an increase in instability resulting from the heating of the surface. But where there's an outbreak of storms, with short lines or clusters and with activity continuing for more than an hour, some force other than heating is likely to be at work. Changes within the air mass result in lifting. Convergence caused by differential heating and the development of air flows as the day wears on could do this. Or a hardly definable front or trough of low pressure at the surface or aloft could be the trigger. Too, the development of one isolated storm could in itself result in a small weather system that would support other storms. Such activity is difficult to forecast with any accuracy, and even after it starts, the forecaster is at a disadvantage. He has to wait to learn the staying and spreading power of the activity to know whether it's an isolated storm or a developing area. The pilot is usually in a better position than the forecaster to develop a good picture. The person on the ground has only sequence reports, satellite pictures, and radar reports to use in making a judgment of development. There's some time lost to this. The pilot has a firsthand view of the building cumulus. If they are numerous and are building upward through the 20,000- to 25,000-foot level, there's likely to be some thunderstorm activity.

NOCTURNAL THUNDERSTORMS

Thunderstorms occurring within an air mass and late at night tend to defy some of the rules of thumb covering thunderstorm development. There's no front, no heating, and no lifting as is found when wind flows over mountains. One explanation lies in upper-level instability being the same, night or day, and convergence caused by the lower-level circulations that develop as a result of temperature changes. This convergence can cause the necessary lifting to trigger storm development. These storms tend to develop so late that they affect only the flight operations of real night owls, and they dissipate by or shortly after dawn. When considered from the surface, they have many of the properties of a warm-frontal storm—that is, the downdraft doesn't penetrate a layer of relatively cooler air near the surface. Aloft, though, a nocturnal thunderstorm can be pretty tough. They aren't always isolated, either. Clusters can develop and cover a relatively wide area, depending on the moisture supply available as feed.

Understanding the theory of thunderstorm development is important to pilots because of the nature of the forecasts we use. To cover every eventuality, a forecaster will often include the "chance" of thunderstorms, which leaves the pilot virtually on his own in making a determination that thunderstorms will or will not be a consideration along the flight path. The basics help us understand why thunderstorms develop and where they are likely to be found. From there, we combine all available information from the ground, from the instrument panel, and from what is seen and felt, to make the important decisions.

2 THUNDERSTORM RESEARCH

While most pilots put maximum effort into avoiding thunderstorms, some seek them and fly through them. Done in the name of research, such operations have contributed greatly to the understanding of thunderstorms both from the standpoint of aviation operations and from the need to be able to pinpoint the type of storm that is most likely to spawn a tornado.

The National Severe Storms Laboratory in Norman, Oklahoma, is a center of thunderstorm research activity. It has two Doppler radar systems which give a new dimension to radar examination of thunderstorms by showing the reflectivity of rainfall in the storms plus motions toward or away from the radar, and areas where motion change is most pronounced. The Doppler radar can thus "see" areas of windshear. (Regular weather radar systems show only the reflectivity of a storm. They lack the Doppler's ability to sense changes in distance from one return to the next and to display motions toward or away from the radar antenna.) Fighter aircraft are used for thunderstorm penetrations; this has been done for a number of years with only one incident of more than minor damage to an airplane. In that case, hail shattered a canopy and battered the leading edges of the airplane. It landed safely.

Storm research at Norman has been going on for a number of years, but recent developments in the Doppler radar systems have added another new dimension to the effort. The radar can be used in real time, to examine and evaluate thunderstorms as they occur. Three values, velocity (movement toward or away from the radar), spectrum width (the measure of change in velocity over a given distance), and reflectivity (as reflected by the rainfall) can be joined together on one radarscope to show the relationship between precipitation and windshear, or turbulence.

Doppler radar can only sense motions toward or away from the radar antenna, so any lateral motion isn't noticed. However, when there is windshear turbulence, air flows are not smooth. There are burbles and churning; as a result, even though the primary motions of the adjoining winds might maintain a constant distance from the radar, meaning that it wouldn't display them, any area of turbulence should show because of the disturbances where the differing areas of wind rub against one another. With the center's two radar units, they can examine most storms from different angles for a better depiction.

The Doppler radar is so sensitive that the development of convective areas can be anticipated before the actual formation of clouds. It will actually track the motions of air. Likewise, the radar can display the motions of air in surface gust fronts ahead of thunderstorms. It can show the relationship between the air flowing into a storm and the air flowing out over the surface. Researchers have concentrated on using this information to better understand low-level turbulence related to the gust front of a thunderstorm, as well as on the continuing study of turbulence within the storm. Use of the Doppler has enabled the lab to add a large base of information to that which previously was based on the relationship between precipitation reflectivity and information provided by actual penetrations by aircraft vectored through storms by the laboratory.

REFLECTIVITY

Reflectivity alone is what we deal with when using airborne or ground radar. There is some difference of opinion among scientists on whether or not reflectivity relates directly to rainfall rate; those at the storm lab do not relate it to rainfall rate. Reflectivity is what the radar senses as it emits energy and displays the result of that energy bouncing off something and returning. The reflectivity of an area of rain is directly related to the size of the drops as well as the shape and number of drops. Drop size tends to be larger in convective activity, and the larger drops tend to be more circular, resulting in higher reflectivity. At times we can see through the rain shaft of a beginning storm even though radar would display it as an area of heavy rain. Maybe it is not heavy to the eye, but it has high reflectivity and any trip through it would bear out the business about drop size. Conversely, we occasionally fly through what seems a rather substantial rain that doesn't show on radar. The drop size was probably small. And while snowflakes can be big, they are not dense and have low reflectivity. Snowfall relates to rainfall on approximately a ten-to-one proportion—10 inches of snow equals 1 inch of rain—which well illustrates the difference in density.

Airborne weather radar sets show reflectivity in three levels of intensity. The highest reflectivity is displayed in red on color sets, and by brighter areas on monochromatic sets. Lower levels are yellow and then green on color radar and lighter shades of green on monochromatic radar. On the latter, this brightest area is blacked out when the radar is in the contour mode, better to identify higher reflectivity within other areas of rain; on some sets this area flashes, alternating between black and the brightest level on return. The word *contour* has been used in this context for years, though it isn't quite correct. The contours would be all the lines defining the three different levels of rainfall, not just the definition of heavy rain. The three levels of rainfall and the gradient between them is what we use to estimate the strength of a storm. Where there is a steep gradient, a change from no rain to heavy rain, in a short distance, that is an indication of an active thunderstorm. Gradient and rainfall rate are the keys on airborne weather sets. The so-called contour feature is there to emphasize heavy rain by flashing it, blacking it out within the other two levels of rainfall displayed, or displaying it in red on a color set.

On traffic radar, areas of high reflectivity are shown by H symbols on computerized radar and simply by bright areas of return on the older narrow-band sets. The level at which the color set turns red, the monochromatic set contours or flashes (or both), or the traffic radar displays H symbols is based on a reflectivity value determined by research to be related to storms that produce significant turbulence. As a guide, most radar sets display precipitation at the heaviest level when the rainfall rate reaches a half inch an hour.

This business about reflectivity has to be taken with a grain of salt, because the distance of a storm from the radar affects reflectivity. What is seen as an area of light return 100 miles away on an airborne set might actually be a severe thunderstorm. Only when the storm is within 30 or 40 miles, or even a bit closer, with no other rainfall between the scope and the storm, and with the radar set's controls set properly, does the picture become representative of what's there. Range from the antenna also affects air traffic control radar's presentation of weather.

Years of research and many penetrations of storms with instrumented aircraft to relate actual turbulence to the radar picture lead to this conclusion: if the reflectivity of any part of the storm is of the highest level shown on the airborne or traffic radar, there is a high probability of encountering severe turbulence somewhere in that storm system. The distribution of turbulence is spread throughout the cloud related to the storm; the greater the reflectivity of the core, the more the turbulence spreads through the cloud associated with the storm. This does not mean that there is one uniform level of turbulence throughout the cloud—there will be big variations. What it means is that the area of maximum reflectivity does not identify the area of greatest turbulence. It can be anywhere; in fact, it is least likely to be found within the heaviest rain. This has been proven time and again by research flights. Also, almost all thunderstorms consist of more than one cell. There are various areas of updraft and downdraft within the storm, which makes it very difficult to pinpoint severe turbulence based on reflectivity alone. (The pilot's visual assessment is no better at picking out smooth spots. Those flying the penetrations have not been able to predict in advance whether the ride will be good or bad based on the appearance of the storm before penetration.)

So when the controller reports heavy precipitation or the airborne weather radar picture is showing the highest level of reflectivity in a storm, the potential for severe turbulence exists in and near clouds associated with the storm. When considering the results of the storm lab's research, bear in mind that they were flying jet fighters and were concerned only with keeping the airplanes away from damaging hail. In other words, if they reported severe turbulence, you can bet that it was severe turbulence.

REFLECTIVITY VERSUS TURBULENCE

The findings about the distribution of turbulence bring home some vital points, especially on the airborne color radar sets that display green, yellow, and red. If a pilot looks at the storm, relates it to the traffic lights he's been looking at all his life, and avoids the red but considers green to mean go and yellow caution, the ride could be bad indeed. If there's red, the potential for severe turbulence exists even in the green. The same goes for areas of lighter shading on a monochromatic set, or the "lightest spot" reported by an air traffic controller.

The ability to see the gust front with Doppler radar has led to research in lower levels, with thunderstorm penetrations flown as low as 400 feet. There's no question about the very strong potential for turbulence in the interaction between air flowing into and out of the storm. Researchers have certainly found this. On into the downdraft, though, it appears that it's not the downward motion of the air that causes accidents during approaches in a thunderstorm. The actual downdraft has been found to be quite mild near the surface. Instead, the trouble comes as the airplane flies from a headwind into a tailwind situation. While passing under a storm this would happen as the airplane passed under the downdraft and started moving away. The rather marked change in wind over a short distance would cause the airspeed to decay and the rate of descent to increase unless the pilot took strong corrective measures.

The observations of downdrafts in the area close to the ground have shown that the down component is only on the order of 6 feet per second, or 360 feet per minute. Again, it is the spread of the downdraft as it converts in wind that causes the problem.

When considering the low-level situation, remember that areas of severe turbulence are not likely to be in the area of heavy rain. There a more nearly steady downward activity is present, with fanning out near the ground. The real bumps come where the air flowing into the storm and the air flowing out of the storm mix.

Researchers hate to generalize, but in pinpointing where shear—burbles and eddies caused by the rub between inflow and outflow—might cause the most severe turbulence, the south or southeast side of the storm is often mentioned. The moisture feed of the storm is usually from the south or the southeast, and that is where the strongest updraft is most likely to be found. However, storms can feed from the southwest, too, and severe turbulence can exist anywhere in the storm.

Perhaps the relationship of turbulence to reflectivity can best be illustrated by what a pilot reports during a storm penetration: "Light turbulence, light precipitation. . . . Precipitation increasing, turbulence remains light. . . . Precipitation very heavy, light turbulence. . . . Precipitation decreasing, light turbulence. . . . Precipitation light, severe turbulence."

IT'S NOT SMOOTHER LOWER

In the recent low-level penetrations operated by the research center at Norman, they haven't found truth in the old wives' tale about lower levels being smoother around thunderstorms. The flow patterns may become a little stronger higher, and the eddies may become larger as you go up, but a significant increase in turbulence has not been found.

There is a wide variety of thunderstorms, and from examining research results it is possible to come to the conclusion that you might be able to fly through a lot of thunderstorms without finding severe turbulence. The majority of storms are small and relatively weak. In some tropical climates you might be able to fly through every thunderstorm of the year without encountering turbulence that would damage an airplane being operated at the proper turbulent-air penetration speed. But remember, the research penetrations are done in fighter aircraft, and the rides are far from smooth.

Geography has a lot to do with the strength of storms. There is no question but that the thunderstorms in the central United States and in northeastern India are the most violent in the world. There are synoptic conditions that spawn strong storms in other areas, but the probability isn't as high as in the two areas mentioned. That's why storm research is done in Norman, Oklahoma. Again, strong is a relative term. As we'll see in the chapter on accidents in and around thunderstorms, they can be pretty tough in other areas. And a pilot doesn't need to find a severe storm to find trouble.

To illustrate differences in thunderstorms, research was done near Singapore to compare a thunderstorm in that tropical area with those around Oklahoma City. The height of the biggest storms proved to be about the same—55,000 to 57,000 feet—but the storms in the Singapore area were in more uniformly moist air. The vertical velocity in the storms was thus quite different, with the Oke City storm the strongest. The tropopause (the level at which the temperature stops decreasing with altitude) is higher in the tropics; as a result, the accelerations necessary to push a storm to a higher level are more even through the storm. When the tropopause is lower, meaning the top of the thunderstorm will reach air that isn't cooling off sooner and instability will be harder to maintain, it takes stronger accelerations to make a tall storm. This helps explain why in tropical areas you can, under some synoptic conditions, go into what appears to be a very heavy thunderstorm and get a lot of rain but no severe turbulence. However, under different synoptic conditions, with fronts or strong lows such as a hurricane, the tropical storm might be as mean as the best Oklahoma has to offer.

A product of some supersonic penetrations of thunderstorms led to a restriction on the Concorde, requiring subsonic flight in any clouds associated with thunderstorms. Originally, the Concorde designers thought that the airplane would be operating above all storms, but research refuted that. They *know* that storms go above 60,000 feet.

HEIGHT VERSUS INTENSITY

What about the correlation between the height of a storm and its intensity? It's always been said that the taller the storm, the meaner the storm, but I had heard that the storm that devastated Wichita Falls, Texas, in 1977 had very low tops. What I heard was not an accurate report. The tops of that storm were up around 55,000 feet. It is true that a tornado can be generated by a storm with lower tops—as low as 25,000 feet—but it wouldn't be a major tornado. Minor tornadoes can occur in relatively innocuous conditions. One tornado report by the public came when there was very little cloudiness around and with the cloud bases up around 10,000 feet and the tops 21,000 feet. The forecaster didn't believe it at first, but radar verified that there was indeed a funnel cloud originating from one of the clouds.

CONTINUING RESEARCH

Research continues to identify what factors lead to severe thunderstorms with the emphasis on how to portray to the individual forecaster how this looks and how he can best communicate it to the public and to aviation interests. In-flight penetrations continue, with an airplane that is instrumented for lightning research. This equipment was added because there had been lightning strikes on almost every flight without instrumentation. Lightning strikes are more prevalent at lower levels; 15,000 feet seems to be the best (or worst) altitude, with the best outside air temperature for lightning between 0° and −5°C. Strikes certainly do occur at higher and lower altitudes; a strike has been recorded as high as 32,000 feet.

How are thunderstorms chosen for penetration?

First, strength is a key. Researchers do not fly any of the small or weak thunderstorms, as they know what is in them; they wait until thunderstorms are at or slightly above the level that would show as a strong cell on airborne weather radar.

The location is also a factor. The researchers would rather fly through storms that are visible to both Doppler radar units, so the storm can be viewed from two angles and an accurate determination can be made of all the air movements within the storm.

Then they look at the storm from the turbulence potential. If the Doppler suggests a turbulent area in the storm, they line the aircraft up and fly through that area, to see what's there. The primary purpose is to verify the radar indications.

When Air Force pilots were flying the program, the only areas avoided were those where the researchers felt there was a possibility of damaging hail, which they define as hail that is three quarters of an inch in diameter or larger. When NASA pilots took over the program, they started culling the most turbulent areas after some minor damage to an airplane.

On the subject of hail size, I was told that quarter-inch hail wouldn't seriously damage a light airplane: "Quarter-of-an-inch hail doesn't have the mass at all. You aren't going that fast. It's a matter of momentum, and usually the smaller hail is softer. You will get erosion with smaller hail; in other words, if you have a fiberglass antenna cover, it can get pitted. It really takes three-quarter-inch hail to be damaging."

CROSSED FINGERS

Researchers don't deny apprehension on thunderstorm research flights, despite 1,500 safe penetrations. But again, these penetrations have all been in fighters. Five-g accelerations have been experienced, and vertical gust velocities have been recorded as high as 65 to 75 feet per second, sometimes as sharp-edged or sudden gusts and sometimes not. If the gusts are 20 feet per second or more the turbulence is moderate; at 30 and above it starts to be severe. In some samples, gusts in the high thirties and low forties were common as well as representative of severe turbulence. This is not unusual in the strong storms that are sought out for penetration.

What about the tales of currents in excess of 250 feet per second in strong thunderstorms? This does happen in updrafts, but it's not precisely the same thing as a vertical gust. The transition zone into such an upward current would be broader than the vertical gusts we've been discussing, and it wouldn't be felt as an instantaneous transition into a vertical current of that strength. The shear area between that updraft and the surrounding air would be broad, though, and incredibly turbulent. To identify numerical values, it is felt that the 66-foot-per-second gust tolerance required of airline aircraft is in a good safe range for all operations. General aviation piston and turboprop aircraft have a vertical gust tolerance of from 30 to 50 feet per second.

In the defining of turbulence, a key is the change in vertical action and the time it takes to change. For example, in one encounter researchers recorded a change from −27 fps to +35 fps in 1.67 seconds. Even wider, from −35 fps to +45 fps in 0.6 second. That would get your undivided attention, even in a highly wing-loaded fighter. When considering those times, also consider that the airplane was moving through at a speed two or three times what the average general aviation aircraft would use for turbulent air penetration. Slower, the transition from one area to the other would take longer.

The worst turbulence is found in the burble, or the disturbed air between the updraft and the downdraft. Again, when taken individually, the drafts aren't so turbulent—it's where they rub together that it all gets mixed up. And that is *not* where the heavy rain falls.

When controllers at the storm lab vector a pilot into a thunderstorm, they give him a heading and altitude to penetrate, plus information on the best way out—right, left, or straight ahead—if it gets to be a bit much and the pilot wants to cut his losses.

An example of how well Doppler radar can pinpoint turbulence involves an armor-plated T-28 that was used for actual hail area penetrations. The researchers were poking at a big storm system with the T-28 plus an F-4. The storm was a distance from home base (near Oklahoma City) and the F-4 got to the storm first, made a couple of passes through, and then returned home. The T-28, being slower, got there as the F-4 was leaving. A couple of penetrations were made without incident, but on the third pass the pilot was northbound through the intensifying line of thunderstorms when the aircraft was zapped by lightning and shaken by severe turbulence. The pilot wanted no more but was on the wrong side of a squall line and didn't have a lot of fuel. The Doppler radar was used to vector him through a smooth area in the squall line, and the event ended happily.

GUST FRONTS AND PULSATIONS

In low-level research, it has been found that the gust front from a thunderstorm system (where the surface wind freshens and shifts as the effect in the downdraft arrives) can extend out as much as 15 or 20 miles ahead of a very strong storm. The top of the gust front is at about 1,500 feet, higher closer to the storm. There would be considerable windshear turbulence at the top of the gust front.

Researchers also find pulsations in storms, and in one example a difference of 32 knots in surface winds emanating from the storm was recorded in a half mile. There would not only be shear at the top of the gust front in this case, there would be shear within the gust front. As a storm pulsates, there can be a series of changes in the strength of the gust front. These changes in wind ahead of a storm system have been related to rain by some meteorologists—where the rain is heaviest, the outflow is the strongest—but researchers at the Severe Storms Lab neither confirm nor deny this because they have not been able to correlate rainfall rates with winds.

ATTENUATION

The lab has done some research on the effect of attenuation on the various types of radar, including the X-Band, which is widely used in general aviation. Attenuation is primarily related to the absorption or reflection of the energy emitted by the radar before it scans its range. This simply means that when you are flying in rain, your radar's capability of depicting weather might be used up in 20 or 10 or even fewer miles. All the energy has been reflected in that distance. Past there, the radar will show nothing, even though there might be a monster storm at the twelve-o'clock position. Even light intermittent rain can obscure a clear picture of what might be ahead, and it's thought that attenuation might be worse at the level where snow is melting into rain.

The researchers took a couple of weather systems as seen by their big S-Band Doppler radar and simulated the picture that would appear on the commonly used X-Band, as well as on C-Band, which has been used by airlines but which requires a larger antenna than would be practical on most general aviation airplanes.

Clearly, C-Band saw less than S-Band, and X-Band saw less than C-Band. This was most dramatically illustrated by the radar depictions of a line of thunderstorms seen from the west side. The storms were moving east, and there was a deep area of rain to the west of the line, between the radar and the really tough weather.

S-Band showed the picture with no attenuation. The general area of rain was clearly visible; the line of strong cells on the eastern edge of the rain area was also clearly visible.

The C-Band radar preserved most of the salient features of the line. The cores were there, but some were not shown at their true value and a few were lost altogether.

With X-Band, the cores and steep rainfall gradients were all missing. Only the general area of rain was shown. The pilot using X-Band would have had to get very close to identify severe weather, and his best information would probably have come from air traffic control radar. However, if the X-Band user was aware of the limitations of the equipment and used it to avoid totally areas of rain in the storm system by a good margin, the radar would have served well. In a way, the wise pilot might well have made the same decision with any of the three radar systems: avoid the whole area.

STORMSCOPE

The lab has used the 3M Ryan Stormscope in some of its research work. This device detects electrical activity and plots the azimuth and approximate distance of the activity on a cathode ray tube. It's not a simple device that tells a pilot that lightning is "over yonder"; the Stormscope processes information to give the most accurate plot possible. And while it is certainly true that electrical discharges are directly related to convective activity, the storm lab has found mature thunderstorms that were not very active electrically. They even found one tornadic storm that produced very little electricity. In another instance, a Stormscope on the ground and a Stormscope in a de Havilland Twin Otter showed very little electrical activity in a storm in which an F-4 pilot found severe turbulence. Conversely, cases were found where there is strong electrical activity but no severe turbulence. Ranging information was also found to be approximate.

These thoughts on X-Band radar and Stormscope have to be taken in context. Neither device has all the answers, and pilots using either should know that. The reservations just expressed come from researchers who use the most sophisticated radar system ever developed for the study of thunderstorm weather. Comparing our lightweight airborne equipment with their banks of computers and huge radar dishes is like comparing a Cessna 210 with a Boeing 747. Just as a 747 pilot might feel that he has mounted a real flivver when flying a 210, the user of an ultrasophisticated radar system is bound to feel that a lightweight X-Band radar or the Stormscope falls short in data presented. The value is in having both our equipment and the knowledge they have gleaned in research with their sophisticated systems.

A lot of positive things can come to aviation users from the work at the National Severe Storms Laboratory. Better displays of radar information can be developed for use at air route traffic control centers and Flight Service Stations. Work can be done to tailor data to emphasize storms that are potentially destructive to airplanes. The current weather radar in use around the country is vintage 1957, and it can and will be replaced with new equipment. Air traffic controllers could have a color display of weather radar data on their scope. This, however, would have to go through the air traffic control computer, which currently lacks the capacity to handle this additional chore. Too, some think that the controller could wind up with too much data. A better solution might be to transmit the picture from sophisticated ground radar systems to the airplane, for display in color on a screen. The picture could be displayed with the aircraft in position, that is, the display of weather would be in relation to the nose of the aircraft as with our present airborne weather radar systems.

In the meanwhile we work with what we have and learn from the research that will lead to better equipment in the future. One thing is certain: all the equipment in the world and all the research won't alter the chaotic nature of the thunderstorm.

3 STORM FORECASTING

Thunderstorm forecasting in the United States is based largely on the product of the National Severe Storms Forecast Center in Kansas City. This office has a number of different programs, the most important of which is identifying areas of potentially severe weather and issuing either tornado or severe-thunderstorm watches. Daily, the center identifies areas where there is a potential for severe weather as well as areas where conditions are favorable for the development of garden variety thunderstorms. The forecast center also develops and issues convective sigmets covering the whole country. These tell of significant thunderstorm activity and are the pilot's number one word on thunderstorms. Convective sigmets are issued hourly. (The word *sigmet* is derived from "significant meteorological advisory.")

The center's work in identifying areas of possible activity by developing a convective outlook for the day is done the night before. By 3:00 A.M. the forecaster has completed the outlook, based on an analysis of the upper-air features at 6:00 the previous evening, plus computer output that develops a model of expected upper-air patterns. Stability, the position of surface highs and lows, and other information is used to make a composite forecast where thunderstorm development is likely. The forecaster is looking at the atmosphere on a broad scale, the work is done well ahead of time, and for flying interests the identification of possible thunderstorm areas in the convective outlook is only a signal to be wary of possible development in these areas. The basic inaccuracies in the computer-generated forecasts of upper winds can, for example, result in large errors in forecasting.

Forecasters around the country use the convective outlook as a guide to forecasting thunderstorms in their area. Certainly if the convective outlook call is for severe activity it will be included in all the terminal and area aviation forecasts. Other potential thunderstorm areas are identified with such a broad brush that, based on later information, a forecaster in the field might or might not include the possibility of thunderstorms on the terminal and area forecasts he prepares.

The convective outlook is a long-term educated guess; a convective sigmet is a statement of fact. It tracks thunderstorm activity that has actually developed and includes a two-hour trend forecast for the activity. In some cases a convective sigmet is issued to cover the expected development of significant thunderstorm activity.

EQUIPMENT

The storm forecast center has a broad array of equipment to use in tracking thunderstorms. Satellite pictures, information from all the weather radar units in the United States, and all other available meteorological information is used. Reports from weather radar units are received at 35 minutes past every hour and are charted. Convective sigmets are issued at 55 after the hour, so the compilers have but 20 minutes to prepare them after receiving the last bit of information. If you get a convective sigmet on the hour, it's based on radar information that is at least 25 minutes old; later in the hour it is even older. It's clear that even this, our most up-to-date information, is but an outline of where activity was located. Real-time information comes only from looking at radar or a Stormscope, by talking to someone who is looking at a weather mapping device, or by visual observation. There's no way sigmets can have pinpoint accuracy, but the forecast center does have the capability of issuing specials and does so when something significant develops unexpectedly between hourly reports.

The storm forecast center identifies areas where severe thunderstorms or tornadoes might occur and issues severe-thunderstorm and tornado watches for these areas. Such watches should be available to pilots. The actual warnings that tell of severe storms or tornadoes that have developed are issued by local weather bureau offices and are not considered aviation products. The FAA does not always distribute these warnings. In other words, an actual tornado or severe-storm sighting report may or may not be available to pilots. The first priority is to the public in the case of severe weather, and the initial warnings go there. This has been roundly criticized following accidents in areas where storm warnings existed and the pilot hadn't been told about them, but there is strong logic to giving public warning precedence. A person on the ground is in a more or less fixed location and is without any other form of information. A pilot should have a broader knowledge of general weather conditions and is in a machine that can easily outrun any storm that might be spotted. In addition, pilots have the use of ground-based air traffic control radar weather capability, many have airborne weather radar or Stormscopes, and convective sigmets tell of thunderstorms, though not on the same dynamic and timely basis as actual reports that lead to tornado warnings or severe-thunderstorm warnings.

CONCENTRATED EFFORT

It might not seem the best deal to have all the nation's thunderstorm forecasting and watching at one place, but actually it is a good way to do it. The meteorologists at the center work with constant input on the subject and continually track existing and potential development.

Satellite pictures are especially helpful. Thunderstorms are quite apparent on the pictures, and when they are run in sequence (they come off every 30 minutes) the development and movement of thunderstorms can be seen rather vividly.

A new system at the forecast center adds "navigational" data to the satellite picture and to radar data. This will precisely identify the location of storms displayed, by latitude and longitude or by their relationship to towns. This has more application to people on the ground, but it does enable the personnel compiling convective sigmets to have very specific knowledge of the location of activity. For the future, an adjunct to this system can, through input from satellite pictures as well as all other sources, refine the winds aloft product to give users and forecasters a much better data base than is currently available. Another refinement of the system can also produce a new and accurate surface chart every 30 minutes. If this new technology becomes fully developed, the quality of forecasting will improve.

The thunderstorm forecasters and the preparers of sigmets rely more on a combination of satellite pictures and teletype data giving the location of thunderstorms than on actual weather radar pictures, which can be called up from around the country and displayed on a screen. The individual radar depictions are perhaps out of context in the big picture that the forecasters have to develop.

SELF-ANALYSIS

The National Weather Service freely admits that it can't predict the location of occurrence of a tornado or the onset of a thunderstorm with accuracy that satisfies the aviation requirement. And the errors in forecasting the time of occurrence of inclement weather are more prevalent than errors in forecasting whether it will or will not occur in a given span of time. This often results in the aviation user perceiving an error in forecasting when the forecast was actually considered accurate. For example, consider an Indianapolis forecast calling for 3,000 overcast and 5 miles visibility from 1500 to 1900, with occasional 500 overcast, one mile, thunderstorm, heavy rain, and wind gusts in excess of 35 knots. If a pilot stopped at Indianapolis for fuel at 1530, there were no thunderstorms around, and he continued along his way without seeing any, he might well think that the forecaster was crying wolf. But storms might develop or arrive in the area at 1845. This would make the forecast basically correct, and the forecaster wouldn't have felt that he needed to amend his forecast to cover the tardiness of the activity. Again, the forecast is only an alert to the possibility of activity. It is up to the pilot to pinpoint the when and where of the matter.

The shotgun approach to forecasting can give the pilot problems when planning a flight.

One early summer day I was headed to the east coast from Cedar Rapids, Iowa. An active cold front had gone through that area early the day before and was moving eastward. It was an unusually strong weather pattern for summertime, and its eastward progress was quite rapid. When I launched at four in the afternoon, the thunderstorm activity was bearing down on the east coast.

The forecast for Trenton, my destination, was for 4,000 scattered, 10,000 overcast, and 2 miles visibility in light rain. Additionally, there was a risk of 800 obscured, half a mile visibility in a heavy thunderstorm, with gusts to 48 knots. This forecast was valid from the time I got it, at 2045Z, through 0300Z, or for the next 6 hours and 15 minutes. The valid time was really longer than that, because it had been in effect for some time when I called. As I looked for an alternate airport, all forecasts on the east coast were similar. But I finally found one, Baltimore, where the forecaster lowered conditions only to 1,000 and 2 in those thunderstorms.

There was a squall line ahead of the cold front. This is where the thunderstorms would be, and they wouldn't affect any one station for more than 45 minutes or an hour. Yet they existed on the forecast for six hours or more, plaguing pilots looking at the "alternate airport" box on the flight plan. An understanding of the weather situation made it possible to do an intelligent job of planning a flight into the area, but it sure didn't help at all with a legal alternate.

One of the reasons forecasting is approximate is that upper-air soundings are made only twice a day and the basic information used in preparing forecasts is about 12 hours old when the forecast is issued. Computers are used to project upper-air patterns that are expected to develop, but nature can be fickle and may not behave as the computer suggests. Too, there's a lot of averaging and smoothing in the computer projections that can hide aberrations in wind at the 500-millibar level. Anyone who has flown a relatively slow airplane at or near the 18,000-foot level very much can tell you that the wind forecasts vary from reasonably close to wildly inaccurate, with the latter the case 30 or 40 percent of the time. This is the level where the patterns aloft affect surface patterns, so it's no wonder basic forecasting misses the mark so often. As mentioned earlier, there is hope for improvement in this area.

Weather systems can speed up or slow down, or new developments can completely change a picture in a relatively short time. One forecaster told me that when the situation is one with a long wave trough—that is, a series of low-pressure troughs at the 18,000-foot level, as shown in Figure 10—the situation will be fairly predictable, with weather systems moving through a given area every two or three days. On the other hand, the straight east–west flow that we generally associate with good weather can start to buckle and develop a trough aloft rather quickly. If such buckling starts right after upper-air soundings are taken, it can lead to the rapid

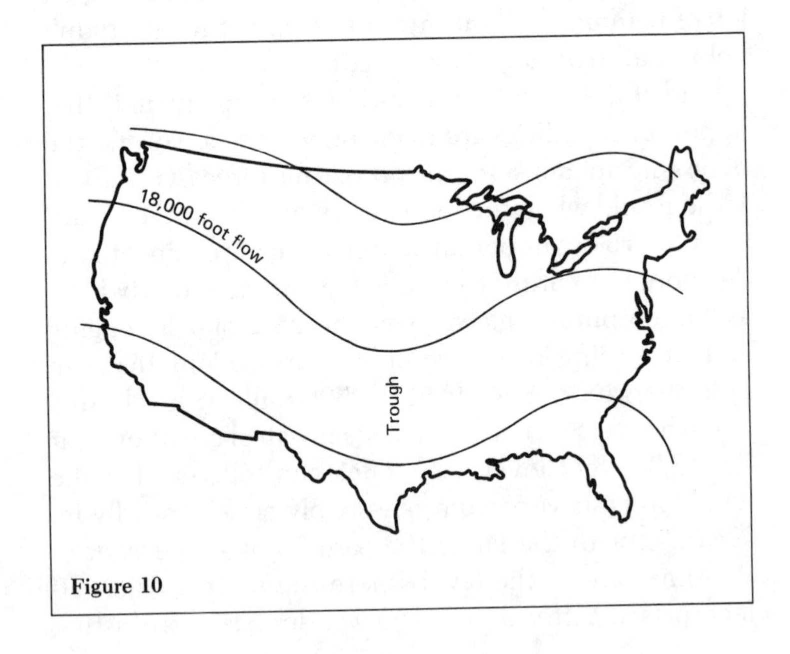

Figure 10

development of a weather system that was waiting for just one more ingredient (instability) to get going—and it can defy the forecasters for a while.

SWINGS AND MISSES

So even with the sophisticated equipment at the storm forecast center, it's tough to catch them all. In fact, only about 30 percent of all tornadoes occur in tornado watch areas. Fifty-five percent of the middle-to-strong level tornadoes do hit in watch areas, and the center has almost a perfect record on the very strong tornadoes that cause widespread property damage. But the number of smaller storms that occur out of watch areas is a good example of how difficult it is for our present forecasting system to anticipate every eventuality.

It is ironic that on the evening before I visited the forecast center there were storms in the Kansas City area that were not covered in the terminal forecast and that did not prompt a convective sigmet until development was well along.

What happened? That morning a rather weak surface low to the west and a low-pressure trough aloft out in western Colorado looked benign. Nothing in the situation appeared strong, and no one anticipated thunderstorm development. The primarily lacking ingredient was low-level moisture. As the day wore on, the surface winds tended to shift from a southerly to a southwesterly direction, which is indicative of less moisture over the central United States—air assumes the properties of the surface over which it flows and Mexico is a lot dryer than the Gulf of Mexico—so the forecasters remained comfortable with the lack of storms in the forecast. That the trough aloft was not expected to move eastward was mirrored in the wind aloft forecast that I got for my trip into Kansas City. It called for straight westerlies. As I flew along, though, I found the wind to be more southwesterly than was forecast. So a trough had developed and there was chatter on the air traffic control frequency about thunderstorms. A line had apparently formed in Kansas and northern Oklahoma with scattered activity in Missouri. As I neared the Kansas City area from the south I could see strong buildups ahead, both visually and on radar. Forecast or not, thunderstorms were on the prowl.

The activity generated heavy rain in the Kansas City area and there was actually one tornado reported but it was never confirmed. The activity developed because the upper-level trough did move eastward and because the surface low strengthened and probably tracked a bit farther south than had been expected. While the low-level moisture supply wasn't strong, it was adequate when instability and lifting became strong. It was a delicate balance, and everything worked for thunderstorms. In other cases, everything looks right and then they don't come because the balance shifts a bit against them.

When strong systems are clearly defined, the forecasters seem to do a reasonable job of forecasting storms. That is not to say that they do well at pinpointing times and places of occurrence even in strong outbreaks. All they can say is that thunderstorm activity is possible during a given time and in a given area.

When using the product of the forecaster, we have to fly with a continuous reminder that weather is dynamic and relate this to the available weather information. Forecasts are based on information at least a half a day old. When the forecast is made, the information input stops. But the weather goes on changing. Maybe it changes as the computer and forecaster thought it would, or maybe the systems decide to do something different. That's why the most important thing that we do with forecasts is question them.

4 THE AIRPLANE AND THE STORM

Many light airplanes lost in thunderstorm areas suffer an airframe failure before reaching the ground. The implication is that the storm "broke" the airplane, but this is not usually the case. In fact, most airframe failures are not directly related to the interaction between the airframe and the turbulence. The more common theme is a loss of control (which is related to turbulence), a very rapid airspeed buildup, and an airframe failure that is related either to pilot-induced overloads at a speed well in excess of the redline or to the speed alone.

Airplane designers do good work on building airframes to withstand turbulence encounters. At a minimum, general aviation airplanes are designed to withstand vertical gusts of the strength found in moderate thunderstorms, and airline aircraft are built to withstand the gusts of all but the most severe storms. The strength, though, is keyed to an airspeed: the airplane will fly through a vertical gust of the specified strength only if it is being operated at the proper speed. Even though that is the case, and even though the airplane is usually within the area affected by thunderstorm turbulence when it breaks up, most of the breakups are still related to extremely high speed or g-loading related to the pilot's efforts to recover from the situation instead of to some moderate airspeed excursions above the correct value. The incidence of airframe failure related solely to an encounter with turbulence is low.

SPEED AND GUSTS

Most general aviation airplanes are designed to withstand a vertical gust of 30 feet per second at maximum cruising speed, the top of the green on the airspeed indicator. The design gust for redline speed is 15 feet per second. Recently designed airplanes, those for which original certification was sought after the early 1970s, are required to withstand a 50-foot-per-second gust at the top of the green and 25 at the redline. This sounds like a big difference between an older and a newer design, but it's not as big as it appears. When the requirement was upped to 50 fps, a change was made in the manner of calculating the onset rate of the gust. While the more recently designed airplane might be stronger in relation to a gust, it's probably only a little stronger—certainly it doesn't hold a five-to-three advantage as suggested by the required gust numbers.

In addition to the FAA requirements, manufacturers have available a body of information from NASA flight data recorders installed in a wide variety of airplanes (including the author's). These devices record speed, g-loading, and altitude on a continuous basis, and the information is considered as indicating the history of speed, g-loading, and altitude relationships in actual operations. Not every airplane carrying a recorder has penetrated the core of a thunderstorm (mine sure hasn't), but enough have to develop a relationship between airplanes and turbulence.

The requirements and the data from flight recorders sets the environment for which airplanes are designed. Mathematical determinations are next used to determine the loads the structure will be subjected to in static test. The wings, fuselage, engine mounts, nacelles, aft fuselage, and horizontal and vertical tail are all considered. In static test, the steady-state load is applied and then the simulated gust loads are added. The loads applied are determined for the most unfavorable center of gravity position. The maneuvering loads are considered separately, as maneuvers load the rear of the tail and wing structure (because they are imposed partly by deflection of the control surfaces) more than gusts, which primarily load from the forward part of the structure.

WEIGHT AND GUSTS

Gust load factors are a function of wing loading—the lower the wing loading the higher the load factor in a given gust. With the airplane at lighter weight, the effect of the gust on that part of the airplane that is always a fixed or near-fixed weight (such as the front fuselage from the pilot's station forward) is more pronounced because a given gust will result in more vertical action.

Because there is more response, or acceleration, found in gusts when the weight is low, and because there is less relieving load in the wings and thus more wing bending when fuel is low, testing is done for zero fuel weight (a limit imposed only on some airplanes—all weight in excess of this value must be in fuel) and for the lightest possible aircraft weight, where accelerations will be the highest.

To develop a mental picture of the effect of aircraft weight and weight distribution on the stresses that will result from an encounter with a vertical gust, look at the airplane from ahead. First, consider it at gross weight with the seats full and the fuel tanks full. Push it upward and the loading is pretty even. The wings are heavy with fuel and the cabin is heavy with people and things. It all goes up together, but the total weight is great and the wings have to withstand this. Next, consider the airplane after much of the fuel has been used out of the wing tanks. The wings are lighter; the fuselage is still the same weight. The total airplane is lighter, too, and a vertical gust will propel it upward at a greater rate. But the weight of the fuselage is relatively higher and will resist the upward shove as before. This puts a bending load on the wings as they make peace with the gust that wants to go up and the fuselage that doesn't. Finally, consider the airplane with little fuel and only two on board, in the front seats. Here the vertical acceleration caused by a gust encounter will be highest, because the airplane is even lighter but the wing bending load won't be as high as with the fuselage full. High stresses might be found in the forward fuselage, though, which is relatively heavy and which has remained at about the same weight in all three examples.

CENTER OF GRAVITY

In addition to weight, the center of gravity position has some effect on the strength of an airplane in a turbulence encounter. When the center of gravity is forward, there is download on the horizontal tail as the airplane flies along normally. As the center of gravity is moved aft, there is less and less downloading on the horizontal tail. It is only logical that the horizontal tail will tolerate less positive loading when the airplane is loaded to its aft center of gravity limit, because there is less offsetting download. Remember, though, that the structural strength requirement is considered at the worst possible condition of weight and center of gravity, so it isn't a case of an airplane being "weak" at any specific loading. It's just stronger than the requirements at the more advantageous conditions of loading. As far as handling qualities go, the airplane is less stable longitudinally with the center of gravity aft. It has less resistance to pitching moments as well as lighter elevator forces. So more control inputs might be required and there could be a greater possibility of overcontrolling with the cg aft than with it forward.

When considering aircraft strength, we tend to look in the pilot's operating handbook at the g-loads the airplane will withstand. The limit load factor for a normal category airplane is 3.8 g, meaning that it will withstand that many times the force of gravity without suffering any damage. (The airplane is at one g in level flight, so 3.8 is an additional 2.8.) For utility category airplanes, the limit positive load factor is 4.4, and it's 6.0 for acrobatic. These limits are applicable to maneuvering loads; the gust criteria are more often the determinant of required strength in an airplane than are the maneuvering loads. The airframe is designed to withstand 150 percent of the limit loads without failing.

In saying that variations in weight change an airplane's response to a gust, we've said that wing loading is part of the equation. The lighter the wing loading, the more response to the gust. An airplane that has 174 square feet of wing and a 4,000-pound gross weight has a wing loading of almost 23 pounds per square foot of wing area. Fly that airplane at 3,000 pounds and the wing loading is just over 17 pounds per square foot, so the airplane will simply respond to the gust with more vigor. The ride will be perceived by both the airframe and the occupants as rougher.

SPEED

Speed is also part of the equation. Just as a lighter weight results in a higher g-loading when encountering a gust of given strength, an increase in speed also results in a higher g-loading. It might not feel the same—hitting the gust at higher speed might result in a jolting ride whereas doing it at light weight gives more of a feeling of being tossed about. In either case, lighter weight or higher speed, the stresses or accelerations are increased. These facts form the basis of what a pilot should do in anticipation of an encounter with potentially severe turbulence: the airplane should be configured in advance to fly at the proper speed.

Even though the airplane's strength in relation to gusts is determined by a requirement related to the airspeed represented by the top of the green arc on the airspeed indicator, this is not the value most often given for turbulent air penetration. The best speed is the speed at which the airplane will stall, and thus relieve the stresses, as the limit load factor is reached. As it relates to maneuvering loads, this is called maneuvering speed, and it is determined by a simple mathematical formula. Multiply the square root of the limit load factor (3.8) by the stalling speed for the clean configuration and presto, that's the maneuvering speed. It can be estimated simply by doubling the number at the bottom of the green arc. Because the technical definition of maneuvering speed relates it to full or abrupt control movements, it might be modified slightly by strength considerations in the aft area of the wing or horizontal tail, or of control surface of system considerations. Gusts, on the other hand, start their work at the leading edge of the wing. The end result is the same—g-loading—only the onset is different. Turbulence is likely to be a series of increases and decreases in g-loading; maneuvering loads are smoother in application.

Actually, the best speed from a structural standpoint for turbulence penetration might be somewhat above maneuvering speed when only the gust is considered. Maneuvering speed is a conservative offering. In turbulence a pilot is likely to be working the controls pretty hard in keeping the wings level—you might even say they are being moved abruptly in many cases—so maneuvering speed as related to full control use might well be the most applicable speed to use. You'd certainly want the option of full or abrupt control travel (especially in the lateral sense) when the airplane is assaulted by the shears and eddies associated with a thunderstorm.

Maneuvering speed gives a good margin above the stall, it gives good protection against the gusts, it gives the pilot the option of abrupt or full control usage, and it is a speed at which the airplane will stall before damage is done. There's a lot in its favor.

Another advantage in maneuvering speed being on the low side is found in the nature of gusts. Doing all this mathematically and on a static test rig is one thing, doing it in the airplane is another. There are horizontal motions with the vertical motions in convective activity, especially at lower altitudes, and a rapid increase in the horizontal component of a gust affecting an airplane can have a dramatic effect on airspeed. Airspeed will increase if the gust value is increasing from ahead, and the airspeed will decrease if the gust velocity is increasing from behind. That's why margins are important. The airspeed jumps around a lot when an airplane is in or near convective activity, and there's little chance that a pilot will keep it pegged on a specific value. So the target value needs some margin, and maneuvering speed gives us the best possible margin from all things. If the airspeed is 20 knots high because of a gust encounter, that doesn't mean the airplane will go to pieces on hitting the next bump, and if it's 20 low it doesn't mean the airplane will automatically fall out of the sky.

What if the speed is low and the airplane encounters a strong vertical gust that increases positive loading to the point that the airplane stalls?

A light airplane is likely to fly on through such a stall with a beep from the stall horn and perhaps a nose-down pitching action. At least the nose-down action is what I've noticed, and other pilots have reported it in severe turbulence encounters that activate the stall warning horn or light. Theoretically, the airplane could enter into a stall and spin, but an airplane with low wing loading and relatively light total weight adapts to any new environment rather quickly. Any stall would last only until the airplane started adjusting to the new situation. A heavier airplane might have more trouble, though there has been no recent history of pilots losing control of jetliners in turbulence.

THE NUMBERS

Airplane manufacturers design extra strength into airplanes to give even more than the FAA-specified strength requirement, which is considered adequate. In testing, the manufacturer is required to go to one and one half times the requirement—a 3.8-g airplane must test to 5.7 g without breaking, although things are allowed to bend—and most go beyond the requirement to see what happens.

Bending before breaking is important. In theory, the structure of the airplane won't be damaged up to 3.8

g if the airplane is operated at the proper speed. Above that g-loading, the airframe might suffer damage, and at least one designer described how his airplane would bend: "We try to have the point where the wing will bend somewhat outboard. It might deform, but you can get home. You end up with wrinkled wings and a very scared customer. Hopefully he can buy one with good wings and try it again." The bending of a wing might conceivably help relieve any turbulence-induced positive loading, thus becoming part of the solution.

Both experience and the examination of accident results convince most people that airplanes lost in thunderstorms don't often get overdoses of g-loading from turbulence. Loss of control is more likely, with very high g-forces imposed on the airplane in a recovery. Or the airplane's speed might go far past the redline and into the flutter regime that is not examined in flight testing. Airplanes are tested beyond the limit speeds shown on the airspeed indicator and in the operating limitations. But in an out-of-control situation, such as a diving spiral with power on, airplanes, especially retractables, will accelerate rather quickly to speeds well in excess of any testing requirement. Flutter, which is something that any airplane would probably encounter if pushed to a high enough speed, is the development of an unstable oscillation in part of the airplane. A horizontal tail surface or wing has some elasticity, and at very high speeds and conditions of loading the aerodynamic forces can interact with the airframe in a manner that can excite the structure and allow flutter. Excessive play in a control surface could contribute, or the flutter could originate in a control surface. Once flutter of a surface starts, the airframe is probably destined to fail in a very short time. There have been cases of pilots encountering flutter at high speeds and coming back to tell about it, but these are rare.

Again, flutter occurs only at a speed well above the limit speeds for any airplane. Pilots don't reach those speeds except when out of control. Aerodynamically clean airplanes accelerate faster when control is lost; the result is that most of the airplanes lost in airframe failures are retractables.

BIGGER = STRONGER?

Pilots often feel that bigger airplanes are "stronger" in turbulence. This isn't necessarily the case. A light single and a pressurized twin are designed to the same gust criteria. From a structural standpoint, if both airplanes are operated at the proper speed, both airplanes should fly through the same turbulence without structural damage. That doesn't mean the rides will be the same. Far from it. The lighter the wing loading, the case with the basic single, the greater the feeling of being tossed about. Speed has something to do with it, too. The proper speed for turbulence penetration on the twin is higher than for the light single. It's going to hustle through the shears, eddies, updrafts, and downdrafts while the slower airplane will spend more time there, giving its pilot extra time to consider that the finest point of thunderstorm flying is absolute avoidance. Too, when the rolling and twisting motion of shear is considered, the physically smaller airplane is more affected because more of the airplane can be contained in any given hunk of churning air. This can also be related to wake turbulence, which is a twisting motion of the air. Put a small airplane behind a big one, and trailing vortices from the big one's wingtips are enough to roll the little one right over. The small airplane can be virtually contained within the vortex, and its control power is simply not adequate to offset the spinning vortex. Put a larger airplane in the same vortex and it might get banged around and the pilot might have to use all the control to avoid an upset, but he can probably manage.

One point on small versus large: a lighter airplane tends to tell the pilot that it's time to slow down when the bumps get bad. The "tossing about" feeling strongly suggests a bit much speed. The heavier an airplane gets, the more a pilot has to rely on the airspeed indicator for information on the airplane's speed in relation to turbulence. It has been my personal observation that if a light airplane offers an exceptionally good ride in turbulence, the pilot has to be more aware of turbulent airspeeds lest he take the good ride as an indication that things are okay.

As far as controllability goes, any airplane has adequate control power to handle turbulence—not to smooth it, to handle it. The wings won't stay level, but when turbulence does displace the airplane laterally, there is control power to arrest the rate of displacement and to get the wings back to level. That sure doesn't mean that an airplane won't come perilously close to being out of control in a thunderstorm. The roll rates can be very high and even a sharp instrument pilot will be hard-pressed after a short time in the turbulent area of a thunderstorm. The feeling, the workload imposed by the airplane wildly lurching and pitching, builds pressure in a hurry. The pilot has to make decisions on control movements quickly and then be instantly ready for the effect of the next disturbance. One wrong move could put the airplane in an even worse situation. But the controllability is there. When we say that a pilot lost control of an airplane, it's seldom because the airplane could not have been controlled; it's usually because the pilot failed to control the airplane. Oh, true, there have been cases where the turbulence or shear was such that the airplane's ability to be controlled was in serious question, but the majority of general aviation airplanes lost to thunderstorms are not lost to the most severe variety.

LIGHTNING AND HAIL

Lightning is an important part of every thunderstorm and, while not a lot of airplanes have been lost to lightning strike, some have.

Design considerations to minimize the potentially destructive effect of lightning strikes include keeping 18 inches of metal outboard of the fuel tanks. The fuel filler caps are kept another 18 inches inboard, or strike-proof caps are used that transfer any arcing to the adjacent skin instead of into the tank. That last 18 inches of metal on the wing (any fiberglass tip is not included) is considered the lightning strike zone. Lightning can strike other extremities as well, but the wing is full of fuel and you sure don't want big sparks running around inside the tank. On a twin, the nacelles are also strike prone, so the guideline about keeping fuel more than

18 inches from the last metal also applies where tanks are included in the nacelle.

Fuel vents are allowed at the tip, but the design must be such that flame won't propagate into the vent system and tank if the tip should be struck, even with raw fuel venting from the tank.

What about tip tanks? Airplanes with tip tanks have fuel closer than 18 inches to the extremity of the wing. This is allowable with a required skin thickness for the tank—enough thickness to act as a conductor to carry away the strike without having a problem in the tank. Cessna twins with tip tanks don't meet the skin thickness requirement (they were certificated before the requirement and are protected by a grandfather clause) and have a lot of history without a case of a lightning strike problem related to tip tanks. The cone on the front and the fairing on the back of the tank have been struck, but they've not had a perforation of a tank resulting from a lightning strike.

Hail, another significant product of thunderstorms, isn't considered in light-airplane certification. On heavier airplanes manufacturers have to show that the engines can accept and digest ice balls without damage, but there's no requirement anywhere that says an airplane has to be able to plow through hail of a given size and survive. There have been cases of both light and heavy airplanes encountering hail, and while I can't recall a case of an airframe being rendered totally unflyable by a hail encounter, engines have been snuffed out, windshields have been broken, and damage to leading edges has been severe.

In sum, the thunderstorm-related design requirements are adequate, and manufacturers generally go beyond the minimum. It's rare for an airframe to fail on first encounter with turbulence, or at any time as long as the speed of the airplane is kept in the proper range. Thunderstorm problems are more likely to be related to a loss of control or, in the case of approach and landing accidents, the effects of windshear on an airplane in the approach configuration. There are exceptions, to be sure—there's probably never been an airplane built that would survive the worst that could be dished out by the meanest thunderstorm. But if pilots work diligently to avoid *all* thunderstorms, then any chance encounters should be with ones far below that "meanest" category. There, structure and controllability are adequate as long as the pilot does his work well.

5 PILOT OR CONTROLLER: WHO IS RESPONSIBLE?

All pilots have heard (and told) war stories about controllers letting them fly into thunderstorms without warning, or about wild rides while being vectored in thunderstorm areas. And there have been lawsuits against the government in relation to accidents in thunderstorm areas. The contention has been that controllers have a responsibility to inform pilots of and keep them away from potentially hazardous weather. The results of such cases can provide insight into how the legal system views our thunderstorm problems. One such case is one in which a Cessna 210 was lost in an area of thunderstorms during an IFR flight one March night. The pilot's survivors sued the government, contending that the controller was at fault because he failed to provide radar vectors to enable the pilot to avoid hazardous weather, failed to provide weather advisories, failed to inform the pilot that other pilots were deviating around weather in the area, and carelessly assumed that an area of heavy precipitation was not hazardous.

The pilot was bound from an airport near Charlotte, North Carolina, to Chattanooga, Tennessee. He checked the weather before takeoff and was told of a sigmet and airmet pertinent to his route of flight covering thunderstorms, possibly in lines and clusters, developing in advance of a cold front. The pilot was given reports on moderate to severe turbulence, showers, and light icing in the Chattanooga area. The radar report described to the pilot showed scattered rainshowers and thundershowers 30 nautical miles northeast of Chattanooga.

The pilot filed a flight plan and told the Flight Service Station specialist that he would ask for radar vectors around the thunderstorms.

A bit more than an hour after takeoff, at 6:48:14 P.M., a controller advised the pilot that he would start picking up some precipitation in about 25 miles if he maintained his present heading. He was further told that if he headed toward Gainesville, Georgia (south of course), and then proceeded west on a vector provided by the controller, he would go south of the precipitation and get to Chattanooga from the southwest. The pilot responded that he would go any way that would keep him out of the precipitation.

At 6:58 the pilot stated that he was in cloud and encountering moderate precipitation. The controller suggested a turn to 230, which the pilot made. A couple of minutes later the pilot reported being in turbulence and asked the controller what he was painting ahead.

The controller reported moderate precipitation, and a minute later the pilot reported that he was "breaking through" the weather. The controller then told him that the weather looked pretty good for about 40 miles but that there appeared to be an area of "real heavy precip" about 50 miles ahead.

At this time the pilot was proceeding directly toward the Rome, Georgia, VOR, but there was a cell between his position and Rome; two controllers discussed the possibility of having to vector the flight to the south of that cell. The pilot asked about Chattanooga weather, which was given him along with a pilot report of severe turbulence and downdrafts from an aircraft approaching Chattanooga from the north.

As the pilot was flying toward the area of heavier weather, several airline jets that had departed Atlanta asked for and received deviations around weather in the area the 210 was approaching. Starting at 7:12 P.M., the following conversation ensued between the pilot and the controller:

> PILOT: Looks like there is pretty heavy lightning off at my one o'clock. Wonder if we could drop down to 6,000 and try to get under that overcast and, uh, wonder if I can proceed, uh, directly on Victor-5 then to Chattanooga?
>
> CONTROLLER: Tell you what, I'm gonna switch you to another frequency, uh, 17R, and, uh, make your request there. I'll tell 'em, uh, of your request. Go now to Atlanta center on 118.9.

The controller who had been handling the flight told the next controller on the interphone line that the pilot wanted to descend and that he wanted to turn and go up Victor-5. The pilot's conversation with the new controller follows:

CONTROLLER: 17R, you can proceed direct Chattanooga, deviate your discretion, north or south of course, descend and maintain 6,000.

PILOT: Alright, sir, we're gonna descend and maintain 6,000. I'd appreciate some help around this precip and we'll proceed direct to Chattanooga. Over.

CONTROLLER: Alright, sir, the only way I can help you around it would be to take you west about 35 miles. You got a solid line; however, aircraft have gone through it, said it's just been light to moderate precip and light turbulence.

PILOT: Alright, sir, uh, we'll proceed direct Chattanooga, be the easiest way, out of, uh, 8,000 for 6,000.

A couple of minutes later the pilot reported that he wasn't picking up Chattanooga very well and asked for vectors. The controller assigned a heading of 330, which sent the aircraft toward the narrowest part of a line of precipitation return showing on his radar screen.

At 7:17 the pilot reported level at 6,000 and requested 4,500 feet in order to get below the clouds. The controller was unable to issue the clearance to a lower altitude at this time. At 7:20 the pilot again requested a lower altitude and reported moderate turbulence and a lot of lightning. The controller was still unable to clear him lower, and about a minute later the following exchange was recorded:

PILOT: Atlanta, this is 17R, heading 330 still look good to Chattanooga?

CONTROLLER: Still looks good, sir; it's taking you right up Victor-5, uh, correction, Victor-51 West, I show you got about a 4 or 5 mile stretch of pretty heavy precip, then you should be breaking out into the clear.

PILOT: Alright, sir we'll maintain 330.

At about 7:24 the pilot called Mayday and said that he had just about gotten torn apart, that there was severe turbulence, and that he was down to 3,500 feet. The controller replied that he was right at the edge (of the weather) and should be breaking out momentarily. The pilot failed to respond to the controller's next transmission, 16 seconds later. The airplane had struck a mountainside, elevation 2,930 feet.

In the legal arguing, the plaintiff held that when the pilot asked if he could proceed direct to Chattanooga on Victor-5, he was asking how the weather appeared along that airway. The defense (the government) maintained that this was a request for a clearance direct to Chattanooga. After that request the controller told of a solid line and relayed the information that he'd have to take the flight 35 miles west to get around the weather. The controller did, though, pass on a very optimistic pilot report, but he was only relaying what someone else had said.

A lot of testimony was devoted to technical expert opinions on the capability of air traffic control radar to display weather. It was concluded that a controller could "sometimes" determine the presence of a thunderstorm with air traffic control radar. And the court found that, based on the weather that existed at the time of the accident, "an air traffic controller would not be able to determine the existence of a thunderstorm by looking at his radar."

The court also noted that the air traffic control manual instructs controllers that "weather assistance" is a third priority item, is not mandatory, and that the controller decides whether or not other duties permit the performance of this service. With these limits in mind, the book says the air traffic controller shall "issue pertinent information on radar-observed weather and suggest radar navigational assistance to avoid these areas. Provide this assistance only when the pilot requests it, whether or not you have previously suggested it. Do not use the word 'turbulence' in describing a condition in connection with weather echoes since radar scopes do not show areas of turbulence."

The court held that whether or not required by manuals, the air traffic controller is required to warn of dangers reasonably apparent to him. And the court found that the services provided to this pilot on this night were reasonable and performed with due care.

The court also found that the pilot, "in exercising his duty as pilot in command, made a conscious decision to attempt to go 'direct Chattanooga'."

Further, the court held that, up to and including the time he made this decision, the pilot had been provided with accurate and timely weather information and was not misled by the weather reports furnished him through air traffic control. At the time of his decision, the pilot was informed by the controller that the only way around the area of weather was to deviate 35 miles, and that other aircraft had penetrated the line reporting light to moderate precipitation and light turbulence. This was, according to the court, in complete accordance with the controller's duty to suggest a vector around the weather and to supplement the intelligence he was receiving from his radarscope with other pilots' actual weather experiences. Throughout this event, the only weather information known to the controllers to be pertinent to the pilot's flight path was conveyed to him. The court's finding was that the controllers never knew, nor could they have been expected to know, of thunderstorms in the pilot's path. In other words, the pilot was held to know everything about the weather surrounding him which was known by the controllers and to have "sufficient information to exercise the options available to him as a pilot to avoid weather." The government won the case.

In another thunderstorm case, the government was sued on the basis that a severe thunderstorm warning issued four minutes before a crash had not been issued to the pilot. In that case, the court held that no negligence was involved in the delay, because the National Weather Service procedures require that such a warning go first to the general public. The court also reasoned: "Predicting the weather is not an exact science. The forecasts or emission of forecasts is a discretionary function." In this accident, the pilot knew that a preceding airplane had enough trouble with the weather to dictate a missed approach, the inclement weather was clearly visible, and the crew had reported seeing the weather.

In still another case, a pilot was continually warned

of severe weather approaching the destination airport, yet he continued. The pilot reported seeing the severe weather on airborne weather radar and had deviated around some cells before heading directly for the airport. The controller told the pilot that he couldn't guarantee that the flight would remain clear of the severe weather if it continued on course; shortly afterward the pilot lost control of the aircraft and it crashed. The government was sued on the basis of the controller's being negligent in allowing the airplane to proceed. The court emphasized that the pilot in command is responsible and that the controller had fulfilled his responsibilities.

The government was not even found completely liable in a case in which a controller failed to tell a pilot on an instrument flight plan about weather that in all likelihood was displayed on the traffic control scope. (The pilot didn't inquire about weather, either.) In this particular instance, the pilot was held equally liable, because he did not meet the recent-experience requirements for instrument flight.

The legal part of it is perhaps not always directly applicable to pilots, because the nature of thunderstorm accidents is such that we do not become directly involved in any lawsuit on the subject. But the findings do teach us that the absolute responsibility of the pilot in command is clearly understood outside the aviation community.

In a way, the judge who presided over the 210 case understood the division of responsibility between pilot and controller better than most pilots do. A lot of pilots will read the exchanges between pilot and controller and feel that the controller might well have borne some responsibility. If that case had been decided by a jury of pilots, it might have come out differently, because pilots would probably feel that this aviator was indeed asking about weather when he inquired about flying Victor-5 to Chattanooga. And some might feel that when the controller passed along other pilots' reports of good conditions he painted a much better picture than actually existed. On the other hand, this pilot continued into worsening conditions and reported seeing lightning. It's common knowledge that lightning occurs in thunderstorms, along with severe turbulence. Yet the pilot continued. He lost the argument with the storm and his survivors lost the argument with the government.

6 ACCIDENTS RELATED TO THUNDERSTORMS

Thunderstorm-related accidents in general aviation airplanes have shown an increase in recent years. And while you might think that the "typical" thunderstorm accident would involve an IFR flight, this isn't the case. Over half the thunderstorm accidents in a recent year involved aircraft on VFR flights.

The VFR scenario repeats: a pilot attempting VFR flight in marginal weather moves into a thunderstorm area and finally into one of the cells in the area. If the pilot is flying a fixed-gear aircraft, he loses control and collides with the ground or just flies into the ground in the poor weather associated with the storm. If the airplane is a retractable and control is lost, there's a good chance the airframe fails before the airplane hits the ground. Almost half the pilots in VFR thunderstorm accidents had instrument ratings but for one reason or another chose not to use their IFR capability on that last flight.

A year's VFR accident activity as related to thunderstorms gives insight into the nature of the problem:

A Cessna T210N was flown VFR into a thunderstorm area. The airplane apparently penetrated a cell, and the private pilot with an instrument rating lost control of the aircraft. An airframe failure followed.

A Piper PA-32 went down in a manner similar to the Cessna just mentioned—airframe failure and all—the primary difference being that the pilot didn't have an instrument rating.

A student pilot on cross-country in a Cessna 152 apparently tried to fly under a thunderstorm and flew into trees.

A Bellanca 17-31 suffered an airframe failure in a thunderstorm area, within eight miles of where a tornado was reported. The private pilot was attempting VFR flight but had an instrument rating.

A Commander piston twin was lost in the ocean in an area of heavy thunderstorms. The flight was VFR, but the commercial pilot was instrument rated.

Another commercial pilot with an instrument rating did the deed in a Cessna 172, in thundery weather in the mountains. The airplane was apparently under control when it hit a mountain.

A private pilot flying a Tri-Pacer went at the thunderstorms with a stacked deck. It was night, and the pilot had allegedly been drinking.

A private pilot lost control of a Cessna 172 shortly after a VFR takeoff in an area of thunderstorms.

A private pilot flying a Piper PA-28 flew into the ground on a night cross-country flight in an area of thunderstorms.

A private pilot flying a Piper Arrow lost control of the airplane during a night VFR flight in or near thunderstorms. An airframe failure followed.

An instrument-rated private pilot on a VFR flight lost control of his Piper Comanche in a thunderstorm, and an airframe failure followed.

Drinking again: a private pilot flying a Cessna 150 at night hit trees while flying low under a thunderstorm. There was evidence the pilot had been drinking.

A Cessna 172 crashed in a thunderstorm area at night after its private pilot lost control of the aircraft.

And a Bonanza hit trees while flying low under a thunderstorm. The pilot had a private certificate and an instrument rating.

In that year's worth of accidents on VFR flights we see a steady pattern of pilots pressing on into thunderstorm areas. In the cases of attempted flight beneath storms there could be instances of strong windshear in and around a storm contributing to the collision with the ground. Or the pilots might have done what pilots do in VFR weather-related accidents in areas of general rain or low ceilings: in an attempt to maintain VFR they fly lower and lower, in reduced visibility, and finally fly into the ground or into obstructions.

There were several night accidents. Even though lightning clearly defines the problem, VFR in thunderstorm areas doesn't work any better at night if a pilot is the least preoccupied with pressing on. A pilot can wind up surrounded, or in clouds he couldn't see coming, because of general darkness.

IFR thunderstorm accidents were not quite as numerous as VFR in this sample year.

Pilots flying IFR might have greater exposure to thunderstorm problems, but they should also have more information available to avoid storms. On the other hand, the VFR pilot's avoidance problem is simple. VFR means staying well clear of all clouds and flying only where visibility is good. If this basic principle is followed, storms are avoided. The IFR pilot's avoidance problem is complicated by the fact that flying in clouds is part of the deal. Human vision, that most reliable storm-avoidance gear, is compromised. Here's what happened IFR in the same year:

> An airline transport pilot climbing to cruise altitude in a Piper PA-28 apparently encountered a thunderstorm. The result was a controlled collision with the ground.
>
> A commercial pilot flying IFR lost control of a Cessna 210N on approach in an area of severe thunderstorms and tornadoes. The airframe did not fail before impact.
>
> A commercial pilot flying a Cessna 182 flew into the ground during a circling approach to an airport while a thunderstorm was in progress.
>
> An airline transport pilot was descending to land in a Cessna 310 and lost control of the aircraft in an area of severe thunderstorms. A funnel cloud was reported near the accident site.
>
> A private instrument pilot lost control of a Moo-

ney 201 in an area of thunderstorms. There was no airframe failure in flight (as of this writing, only one such failure has been recorded in a metal-wing M-20 series Mooney).

Another private instrument pilot lost control of a Cessna T210 and crashed in a thunderstorm area. There was no airframe failure in flight.

Still another private instrument pilot managed to fail the airframe of a Cessna T210 after losing control in a thunderstorm while en route.

A private pilot who did not have an instrument rating was on an IFR flight plan in a Piper Navajo when control was lost. There was no airframe failure.

Witnesses saw a Bonanza come out of a thunderstorm after the commercial pilot lost control and after the airframe failed.

And an airline transport pilot lost control of a Cessna 320 and crashed shortly after takeoff in a thunderstorm area.

All these accidents occurred on IFR flight plans, and all the pilots save the one in the Navajo were instrument-rated. There was more mention of truly severe weather in the area of the accident than with the VFR flights, and more of the IFR problems came up during the descent or approach phase of flight.

That year the biggest months for thunderstorm accidents were April, July, and October. And when a strong storm system moves across the country in the spring or fall, it usually claims more than one airplane. Likewise, when there are thunderstorms during a holiday period, there's usually more than one accident. The Fourth of

July often stands out as the worst thunderstorm accident period.

Thunderstorm accidents happen in all parts of the country, and while Oklahoma and Kansas might be considered home base for severe thunderstorm activity, there's no concentration of accidents there. In fact, on lists of thunderstorm accidents, the southeastern United States (excluding Florida) and the Ohio valley area often stand out. Storms in those areas might not often be as severe as those that rumble across the Great Plains, but they are adequate to overwhelm the average general aviation pilot. They are also often harder to see and avoid. The really big and bad stuff farther west becomes rather conspicuous when fully developed.

From that summary of activity the next step is to a more detailed study of some thunderstorm accidents. From these we can get a feel for the events leading up to the penetration of a thunderstorm and the problems that develop once the airplane is in the storm.

Many of the accidents examined in detail are air carrier accidents. These are useful in the study of thunderstorms because of the information provided by cockpit and flight recorders.

Air carrier pilots do seem to be learning more about thunderstorms than their general aviation counterparts. Airline involvement in thunderstorm accidents has declined in recent years. That decline has not come because airline pilots have discovered the magic of how to penetrate storms. It has come because of a refinement of thunderstorm avoidance procedures and adherence to those procedures. It is likely that a large air carrier airplane can survive the gust loadings of a thunderstorm better than a light airplane, and it is likely that a large air carrier airplane can be controlled in turbulence that would make a light airplane marginally manageable at best. But size does not make air carrier airplanes immune to damage or to loss of control. The airplanes in the following accident analyses range in size from four-place singles to a Boeing 747.

Midsummer Maelstrom Versus VFR Arrow

A rather weak cold front lay to the west as the private pilot checked weather for a late July flight from Montgomery, Alabama, to the Dallas area. Scattered thunderstorms were forecast along the route. In midafternoon a sigmet was issued for thunderstorms in northeast Louisiana and central Mississippi. A 10-mile-wide line of thunderstorms was described as extending from 30 miles northeast of Monroe, Louisiana, eastward to about 40 miles south of Meridian, Mississippi. That this activity was thought to be spawned by the heating of a July day was reflected in the sigmet: "Activity expected to decrease slowly. Cancel at 1900." The general situation showed stable air up to about 3,400 feet m.s.l. and unstable air above that. The moisture supply was apparently plentiful; heating was triggering the storms.

The pilot was in the Montgomery Flight Service Station for his briefing early in the afternoon. There he was told of the scattered to broken clouds at 3,000 feet with somewhat worse weather in Texas and with occasional thunderstorms along the route. He returned about an hour later for an update and was given a sigmet and was shown a radar facsimile chart depicting a line of cells starting about 40 miles west of Meridian. The FSS specialist updated the en route weather for the pilot and advised him to check weather along the way. Because of the building thunderstorms he also gave him an approach control frequency for the area near Meridian and suggested that the pilot use that frequency for radar service while passing through the area.

The pilot, who did not have an instrument rating, filed a VFR flight plan, specifying a route over Meridian and westward. It was almost four in the afternoon when the aircraft departed from Montgomery.

Meridian was reporting 2,500 scattered, 6,000 broken, 25,000 overcast, visibility 7 miles in light rain at four. The surface wind was southerly at 15 knots with gusts to 27. This surface wind was much stronger than the general surface wind in the area, and the surface temperature was about 10 degrees lower than it had been. Both would be expected in an area affected by the downdraft of a nearby thunderstorm or with a thunderstorm dissipating in the area.

To the northeast, Meridian Naval Air Station's report was 2,500 scattered, 12,000 scattered, 25,000 broken, 7 miles visibility. The temperature was high and the surface winds light. Rainshowers were reported to the southwest, with towering cumulus through west through northwest.

The Arrow's flight plan would take it through this area, and at about 4:25 the pilot contacted the approach control facility at Meridian and asked, "I wonder if you could tell us, any weather you might have between your station and Longview."

The controller replied, "Have a weather area overlying the Meridian Vortac this time, beginning 10 miles north of the Vortac extending southeastbound for 40 miles." The Arrow was 37 miles east of Meridian, flying at 4,500 feet.

At 4:28 the pilot was advised to contact another controller for further flight-following service, and when communication was established the pilot again asked for weather information. The controller responded, "Presently showing a heavy weather area approximately 40 miles northeast of Meridian Vortac extending approximately 20 miles southwest, correction southeast, of Meridian Vortac, and appears to be a heavy-weather area in the vicinity of Meridian Vortac extending northward approximately 10 miles and eastward approximately 10 miles."

The pilot said that he saw the weather and would start going around it. The controller then added that he was showing another weather area directly ahead of the flight, 8 miles away. The pilot acknowledged receipt of this information.

At 4:37, in response to a query from the controller, the pilot reported that he was still flying at 4,500 feet. Five minutes after that the pilot asked for a position report; the controller responded that the flight was 28 miles northeast of the Meridian Vortac. This indicated a northerly deviation from course, and the following dialogue came after the controller asked the pilot if he was going to go on toward Shreveport:

"How's it look between here and there?"

"Okay, show you on the northeast edge of a little line now. After you go about, say, 10 more miles, looks like you could cut back to the southwest."

"Okay, we'll keep giving it a try then."

Three minutes later the sound of an ELT was heard along with a transmission. The transmission was not entirely clear, but the following represents the best interpretation possible under the circumstances: "Seven seventy-eight is, Mayday, 778 is twisting, lost all . . ." The transmission was also interpreted as "Seven seventy-eight, Mayday, 778 [crashing], lost both . . ." The aircraft crashed after breaking up in flight. The location of the accident was about 35 miles north of Meridian, or 35 miles north of the route the pilot had initially planned to fly on his east-to-west course.

Both outer wing panels failed upward and both sides of the stabilator failed due to down loading. Such failures are normally associated with pilot-induced overloads.

In a weather observation taken immediately after the crash, Meridian NAS (20 miles southeast of the accident site) reported 900 scattered, 2,000 scattered, measured 3,300 broken 12,000 overcast, 6 miles visibility in light rain with cumulonimbus southwest through northeast and rainshowers of unknown intensity west-southwest through northwest.

The pilot never reported changing from a cruising level of 4,500 feet. If indeed he never actually changed altitude, this would have put him substantially above the bases of the broken clouds that Meridian NAS reported at 3,300 feet. This also roughly corresponds with the level (3,400 feet) where the air became unstable. Building cumulus would naturally be expected from this level upward.

Whenever thunderstorms develop in lines or large clusters within an air mass, as was the case here, some specific lifting effect or instability that is peculiar to an area is involved in the development of the storms. In this instance, it is likely that a cooler pocket of air aloft increased instability in the particular area of the storms and supported the upward development of the cumulus clouds. Such aberrations in temperature aloft are small in scale and can't be forecast accurately. However, on this day the forecast did call for thunderstorms along the route and the sigmet was effective in pinpointing storms that actually developed.

This pilot was attempting to fly VFR around the north edge of activity. From the descriptions of the situation given the pilot as it appeared on traffic control radar, he was effective in reaching the north side of the line as it was painted on radar. But the aircraft was above the bases of the broken clouds, which complicated the job of maintaining VFR. And as the aircraft neared the area where thunderstorms were developing, it is likely that the clouds became more congested. That the lower clouds changed from scattered to broken as the storms approached was evidence of this. While middle-level winds were relatively light and the storms were not moving rapidly, they were moving from southwest to northeast. That put the flight in the direction of movement. Finally, where there is a short line of storms, a likely area of development is at the end of the line.

There was no indication that this pilot actually flew into a thunderstorm. He could have, but it also appears possible that he flew VFR into an area of increasing clouds, at an altitude where he would have to dodge those clouds to remain VFR. The pilot could have flown into turbulent cumulus and could have lost control of the aircraft in a hasty effort to escape. The aircraft broke up in flight because of pilot-induced loads. It was on a southwesterly heading, and the pilot did not express any concern to the controller before his Mayday message.

In VFR operations, staying out of all clouds is important. When those clouds are turbulent, as cumulus near thunderstorms are prone to be, it becomes even more important. Flying above the bases of the cumulus is certainly more comfortable on a hot July day such as this one, but when the cu start getting closer together, a better altitude is one below the clouds. And if that doesn't give good clearance above terrain and obstructions as well as affording good visibility, the best altitude would be that of the parking area of the closest airport.

May Morning Versus A Turbo Commander

The 7:00 A.M. surface chart showed a low to the west of Oklahoma City, and by midmorning the Oke City weather was wet and thundery. At 9:55, Wiley Post Airport reported 5,000 overcast, 5 miles visibility, and a thunderstorm southwest moving northeast. Lightning in cloud, cloud to cloud, and cloud to ground west was observed. Weather radar was depicting a general area of thunderstorms and rain all around Oklahoma City. It was described as three-tenths coverage of strong echoes containing thunderstorms with heavy rain and three-tenths coverage of light rain with no change in intensity. The maximum echo top was later reported to be 42,000 feet at 297 degrees and 37 nautical miles. Surface winds were light and easterly, reflecting the circulation around and into the surface low. Winds at the 18,000-foot level were southwesterly, with a trough aloft to the west of Oklahoma City. According to the early-morning upper air sounding, the air was relatively stable up to 9,000 or 10,000 feet, where it became unstable. The forecast called for moderate showers and thunderstorms during the morning and intensifying activity in the afternoon as the surface heated.

There is no record of the experienced pilot's getting a weather briefing for a flight to Colorado, but the situation around Oklahoma City was quite apparent at the time of his departure. The thunderstorm that had been reported to the southwest of Wiley Post was, true to form, moving with the middle-level winds and toward the airport.

Air traffic control radar was also painting the weather and controllers were adjusting traffic flow as requested by pilots to avoid areas of weather.

The turboprop Commander was westbound, to Colorado Springs. The pilot had filed for flight level 260 (26,000 feet), and before takeoff he asked the tower to advise the departure controller that he would "like vectors around any thunderstorms that look real heavy to 'em."

The following exchange between the departure controller and the tower controller shows that the pilot's desire to avoid the weather was anticipated:

"Hey, Wiley, ask that 99F if he can take a right turn immediately after takeoff to 260."

"Okay, he wanted me to ask you for vectors around any thunderstorms out there. Will that get him around them?"

"Well, if he can make an immediate right turn to 260, that'll . . . but you got one three southwest of the airport moving northeast. That's about the only way I can get him around it right now."

The tower asked the pilot if he could take that immediate right turn to 260, the pilot responded affirmatively, and the tower cleared the flight for takeoff. The controller added, "And he said they do have a thunderstorm that's 3 miles southwest of the airport and that will get you around it."

The Commander was off at about 10:00 A.M. Clearance to climb to 11,000 was issued and the departure controller instructed the pilot to fly the 260-degree heading until joining the airway. Shortly afterward the flight was cleared on up to FL230 and control of the flight was handed off to the Fort Worth air route traffic control center. At the time of the handoff, the departure and center controllers had some discussion about weather, and about this Commander deviating to the north of the weather. It was about six minutes after takeoff when the pilot checked in with Fort Worth center and reported out of 8,900 feet and climbing. Less than a minute later the pilot said: "And, uh, Center

99F, we'd like to come on around, uh, uh, about, uh, 10 degrees to the north here to miss a shower we are showing on radar."

The center approved that request and subsequently lost both radar and radio contact with the airplane. One controller said, "We saw him out of 10,000 out there and that was the last time we saw him."

The target disappeared to the north of a weather echo shown on traffic control radar. Interpolation of an air traffic control radar plot of the aircraft track and weather radar information suggests that the flight's last location was near the thunderstorm with the highest tops—42,000 feet at 297 degrees and 37 miles, as described in a special radar report compiled 20 minutes after the aircraft dropped off the controller's screen.

Witnesses on the ground in the area reported lightning and heavy rain to the west. The airplane was reported by one witness as being in a right turn (which would have been away from the storm). A man driving along a highway said, "We saw two balls of fire (like Roman candles) and one big object. The fire was out before it hit the ground."

The National Transportation Safety Board's probable cause cited the pilot for inadequate preflight preparation, planning, or both, for initiating flight into adverse weather conditions, and for exceeding the design stress limits of the aircraft. The wings and the horizontal stabilizers failed in flight, and aircraft wreckage was spread over an area about 3,150 feet long and 1,000 feet wide, on an easterly heading. The aircraft had been on a westerly heading as it approached the storm. The examination of aircraft systems showed that continuity of components existed prior to the in-flight breakup of the structure.

The witness who saw the airplane in a right turn included the following in his statement: "I saw the airplane flying in a northwesterly heading. It was attempting to make a right-hand turn. Lightning was very heavy to the west of the aircraft as the airplane was in the right-hand turn. It exploded in midair and went into flames and spun approximately 300 feet, and while in flames it slid off to the southwest. . . . I would also guess the altitude of the plane about 3,500 feet at the time of the explosion."

In a radio transmission, the pilot mentioned weather ahead and asked for a 10-degree turn to the right. The information available from weather radar indicates that the weather was indeed better to the right, so this turn was in a logical direction.

Airframe failure in flight can be caused by an encounter with turbulence at a speed above maneuvering speed, or it can be caused by the pilot overstressing the airframe. The latter is the more common cause of failure and usually comes after first a lateral and then a longitudinal loss of control. For whatever reason—spatial disorientation or turbulence—the airplane is allowed to bank steeply. Or the steep bank might come as the pilot recognizes a bad situation ahead and rolls rapidly into a turn as the aircraft enters turbulence. The bank steepens, the nose drops, and airspeed builds beyond the redline, never-exceed speed. Failure usually comes as the pilot attempts to recover from the diving turn.

If the witness who saw the airplane heading northwest, and then turning, and then exploding at an estimated altitude of 3,500 feet, constructed an accurate picture from his sighting, it is possible that this pilot lost control in a right turn away from the area of weather. The aircraft appears to have turned 90 or more degrees to the right and to have lost a substantial amount of altitude before breaking up.

While no severe weather was forecast, and the thunderstorms were to have been "moderate" in the morning hours, the pilot flying this relatively heavy, radar-equipped turboprop could have been prompted to make a quick right turn by turbulence. The aircraft was to the northeast of the largest thunderstorm cell in the area, which would have been a likely area of turbulence. The airplane was at about the level where instability became strong: this could have combined with the strong feed into the storm as it moved northeastward to create turbulence that was well clear of the precipitation area painted on the pilot's weather radar, and that was unexpected. The fact that the airplane was a good distance from the weather return showed on traffic radar supports this theory. Also, the activity was clustered, and the possibility exists that the pilot could have moved into a developing but not mature cell, forming as a result of the downdraft from that tall cell displacing warm air near the surface and contributing to localized low-level instability.

Strong Shear

The weather in Denver was unstable on the August afternoon of this accident, with scattered thunderstorms and rainshowers in the area. At the airport, the 3:51 P.M. weather observation reported high ceilings and excellent visibility but added, "Thunderstorm ended at 1550, moved east, cumulonimbus in all quadrants moving east, rainshowers of unknown intensity east through south, peak wind 320 at 28 knots at 1519, rain began at 1520 and ended at 1540." A few minutes later, at 1606, weather radar showed a weak small echo about 3 miles in diameter near the Denver Stapleton airport; this grew to a large but still weak echo over the next few minutes and then dissipated.

At 1605 a Boeing 727 air carrier flight was cleared for takeoff on Runway 35L. The tower gave surface winds of 250 degrees at 15 knots with gusts to 22 when clearing this aircraft for takeoff. After liftoff, the pilot reported, "Okay, you got some pretty good up and downdrafts out here from two, three hundred feet."

At 1607 the tower cleared a Convair 580 (turboprop) airline aircraft for takeoff on the same runway. At this time the tower reported the surface wind from 280 degrees at 13 knots with gusts to 22 and gave the crew the previous pilot's report about updrafts and downdrafts. After departure this pilot reported, "There's a pretty good shear line there about halfway down 35." The tower asked for an altitude on the shear and the pilot responded, "Oh about just like that other airplane called it, about 200 feet."

The next aircraft to go was another Boeing 727, which was loaded to near maximum weight for the conditions and the runway. The crew of this aircraft acknowledged hearing the Convair 580 pilot's report.

At 1610 the controller cleared this 727 for takeoff, gave the surface winds as 230 degrees at 12 knots, and added, "There have been reports of pretty stout up and downdrafts and that shear out there at 200 to 300 feet." The flight acknowledged.

The takeoff roll was normal, and when the captain called "Rotate," the first officer (who was flying) brought the pitch attitude of the aircraft up to between 13 and 15 degrees. The aircraft lifted off shortly afterward, and when the first officer saw a positive rate of climb he called for landing gear retraction. The captain said that the aircraft entered heavy rain at about the time of rotation; he turned the windshield wipers on and, on command, retracted the landing gear.

The aircraft climbed normally to 150 or 200 feet, according to the flightcrew, with the airspeed at V_2, the takeoff safety speed, plus 5 knots, or about 148 knots in this case. The airspeed fluctuated and then dropped by 10 knots; in response the first officer lowered the nose. The captain felt the aircraft sink and saw the airspeed decrease another 15 knots, at which time he took control, advanced power to maximum thrust, and adjusted the pitch angle to 10 degrees. The aircraft descended, the captain attempted to increase the pitch attitude, the stall warning sounded, and the aircraft struck the ground on the right shoulder of the runway, near the end.

This takeoff occurred as the small echo observed on weather radar was increasing to a large but rather weak echo. A construction worker in a trailer half a mile from the accident site reported rain and strong winds beginning about 15 minutes before the aircraft crashed. A bit later he heard a loud noise, opened the door, and observed that the roof had blown off a shed north of his location.

An aircraft mechanic in the area said that winds were gusting from the south; he estimated that wind speed varied from near calm to 50 or 60 mph.

Another construction worker said that five minutes before the accident a southerly wind blew sand so hard he took shelter. Then the wind shifted back to northeast.

A third construction worker saw the aircraft descend to the ground and estimated that the wind in the area was from the southeast at 30 to 40 mph.

According to the downed 727's flight recorder, the airspeed was 154 knots when the first officer called for gear retraction. Two seconds later it was 157 knots, and then, in a span of only five seconds, the airspeed decreased from 157 to 116 knots. The crash came 6.6 seconds later, at an airspeed of 126 knots. The flight recorders of the two preceding airplanes were also examined. The Convair's showed a decrease from 155 to 119 knots in 10.8 seconds, starting 17 seconds after liftoff; the other 727's airspeed decreased from 157 to 134 knots in 15.6 seconds. Clearly they had encountered shear, but nothing of the magnitude that greeted the 727 that crashed.

During the investigation, studies were made to determine the direction and velocity of the wind resulting from the outflow of the thunderstorm cells in the area. The effect on the airplane, as shown on the flight recorder, and surface wind observations were used in the calculations. Possible horizontal and vertical wind components were determined for six different performance situations (to allow for variations in the rate of thrust application and other variables), and it was estimated that in its last 15 seconds of flight the airplane transitioned from an area with a 10-to-20 knot headwind to an area where there was a tailwind component of from 60 to as high as 90 knots. The conclusions on vertical component were not as clearly defined, though in every case they reflected updraft and downdraft activity in the last seconds of flight. At impact, the estimates in the NTSB report on the accident ranged from a 5-foot-per-second updraft to a 26-foot-per-second downdraft.

The weather radar report indicated that the thunderstorm dissipated shortly after the accident. Even before it dissipated the storm wasn't readily visible to pilots or controllers, because its base was high above the ground and it was surrounded by other clouds and high-level thunderstorms. The low relative humidity in the area caused much of the rain to evaporate before it reached the ground, and the NTSB speculated that the evaporative cooling of the descending air enhanced the strength of the downdraft at lower levels.

This strong wind was a very localized and variable condition.

When the 727 started its takeoff roll, the outflow from the storm was affecting only the northern portion of the runway. The controller's surface wind information came from an anemometer located southeast of the runway, and while the information he gave the crew was accurate, it was not really reflective of the situation all along the departure runway. The crew did have the benefit of information from the two aircraft that departed before them, but these reports contained only general statements, and those aircraft flew on through the situation with no alarming reports of difficulty.

In its analysis of the accident, the NTSB concluded that once in the windshear situation, the crew could not have avoided the accident. The airplane simply didn't have the performance to fly successfully through such a strong wind shift area. An increasing tailwind always causes a drop in airspeed; an increase of as much as 90 knots was simply too much.

Fortunately, everyone survived this accident, which had an interesting aftermath. After the aircraft stopped, the flightcrew departed from the flight deck through the side windows with the engines still running and developing high power. The captain returned to the cockpit and made an unsuccessful attempt to shut the engines down. A third attempt also failed. The fire department finally accomplished the shutdown by squirting water and foam into the engines. The aircraft was substantially damaged, and the engine controls were apparently inoperative. There was no fire, because all fuel tanks and lines remained intact during the airplane's belly slide after impact.

While this accident was thunderstorm-related, there wasn't really any "thunderstorm weather" as such: no heavy rainshafts and no boiling low clouds. This makes the lesson more difficult. With good visibility and a shifty west wind at reasonable velocities reported by the tower, most pilots would launch even with pilot reports of windshear. In this case it didn't work. Fortunately, the runway was long and the area where the airplane impacted and stopped was relatively flat.

This accident is a clear example of how flying from a headwind or a no-wind situation into a strong tailwind situation can affect an airplane's ability to fly on. The wind shift in this case mandated that the airplane crash. And while you wouldn't often find it so strong, wind shift situations will be found when flying under any thunderstorm.

The message is, beware all thunderstorms. Even the most innocuous in appearance can have quite a punch.

Hazy Summer Day

It was a typical hazy summer day along the middle Atlantic coast. A lazy cold front extended through central North Carolina, with a weak surface low over western South Carolina. Nobody was expecting big weather problems when the instrument-rated private pilot checked weather at Savannah for his northbound flight in a Piper Arrow. Scattered thunderstorms were mentioned for the afternoon, but they are a familiar part of summertime forecasts for the southeastern United States. The pilot had filed IFR for previous legs of this extended trip, and did so for the flight from Savannah. The plan was to fly up Victor-1 at 9,000 feet. As events unfolded, this pilot's experience gives a dramatic example of the limitations of weather reporting and the limitations of air traffic control in thunderstorm situations.

When the pilot was briefed at Savannah, no convective sigmets were current for the route. A few minutes after takeoff, departure control advised all pilots on the frequency of a convective sigmet and outlined the area affected. The pilot may or may not have switched to the VOR and listened to the complete sigmet when it was broadcast. Even if he had listened, it probably wouldn't have influenced the decision to continue the flight. The sigmet wasn't that foreboding: it advised that there were a few intense thunderstorms and scattered less intense thunderstorms west of a line between Florence, South Carolina, and a point 20 miles south of Macon, Georgia. This area lay to the west of the route of flight. Other sigmets were subsequently issued for thunderstorms along the route of flight, but the pilot was not informed of them. Even if he had been, they would have told him only of scattered activity.

Maybe it was scattered in the beginning, but the thunderstorm activity that would ultimately affect this flight was building rapidly and expanding in airspace controlled by the Washington air route traffic control center. This center was staffed with meteorologists whose duties included the evaluation of weather data with a responsibility to advise controllers, through supervisors, of hazardous weather in their sectors. The meteorologists did not at this time have information from the weather radar system at Wilmington, North Carolina, though they probably did have access to the traffic control radar. Whatever the case, they played no role.

After leaving Savannah's area, the aircraft was under the control of Jacksonville center. The first mention of weather, or thunderstorms, came at five after four in the afternoon: "Ah, 786, on my present route of flight am I, ah, gonna be goin' through thunderstorms?"

The center controller replied, "41786, uh, I'm painting some thunderstorms both, uh, to the east and west of the airway. You can deviate as necessary for weather avoidance. Back on course when you can. You should be able to pick your way through. We've had about three or four of them get through that area here in the last, ah, half hour."

In response to that, the pilot requested "help vectoring."

Apparently things progressed smoothly until 4:52. The controllers offered no vectoring and the pilot voiced no complaints about the ride or concern about the view ahead. The general cloud situation suggested that the pilot should have been on top most of the time, with visibility restricted by typical summertime haze.

At 4:52, control of the flight was handed off to Washington center, where it was apparent that thunderstorms were becoming a problem. The pilot of a Bonanza reported severe turbulence, that he was in a cell, and that his altitude was varying. At 4:51, before the Arrow pilot tuned into the frequency, the controller told the Bonanza pilot, in relation to a gap in the weather, "That hole is closing up now. If you'd like, uh, it might be best just to land Myrtle, there for a while, 'cause looks like it's building pretty bad." A minute later the controller said to the Bonanza pilot: "Maintain your 310 heading, now, we'll take you up through a hole up there. I think we might be able to put you through it."

At 4:53, the pilot of the Arrow checked in with Washington center, level at 9,000 feet, and asked if there was any weather up ahead. The controller didn't reply for a moment. He was busy vectoring the Bonanza. When he did get to the Arrow pilot he said, in reply to the question about weather: "That's affirmative. Uh, you might have to, uh, go around this cell at your—off your left hand side, then find a hole that's up. I got another aircraft descending through, and then, uh, go up towards Fayetteville, that way. But, uh, going over Wilmington looks like it's, uh, be hard."

The pilot replied with a request for vectors and a statement that his aircraft was not equipped with radar. A few minutes later the Bonanza reported more turbulence and a descent rate of 4,000 feet per minute. Presumably the Arrow pilot heard this.

At this point, the controller who had been working was relieved by another controller. He said that he took over because he was concerned that aircraft in the sector were not being adequately vectored around weather. He also said that he switched from narrow-band radar depiction (where information is run through a computer and then displayed on the scope) to broadband (a direct display of aircraft and weather returns) and started working with aircraft already in the weather. This controller said he was not advised of the Arrow pilot's request for vectors.

His first instruction was to a third airplane on the frequency (in addition to the Arrow and the Bonanza): "Okay, uh, uh, 943 Tango, I would recommend a 120 heading for about 20 miles and then direct Fayetteville. There is a slight hole—let me check it on my broad band—okay, there's a hole in your ten o'clock and about, uh, four miles. You go straight, almost a 270 heading for about 8 miles. According to my radar, that'll take you right through the hole and then direct

Fayetteville should be a clean shot if you don't, uh, take too much time."

The Bonanza came next, with more discussion about weather. At five o'clock, 943 Tango had another question. Then this controller had a rather lengthy exchange with the Bonanza pilot that included the following transmission from the controller to the pilot of the Bonanza: "Okay, be advised I do not have weather radar. I can take you to an airport that's clear of it which would be Florence or, uh, Myrtle Beach to the southeast. You gonna have a tough time getting north unless you're visual. I got a solid line extending from Wilmington all the way up to, uh, Fort Mill with occasional holes in it. There's one good hole that may be there, according to my radar, at twelve o'clock and ten to twelve miles. However, I do not have weather radar."

At this time the Bonanza pilot said that he didn't have weather radar either and would go to the closest airport: "If you give me the, uh, best-looking picture you got, that's where I want to go." (The Bonanza pilot later said that he had been in the clear at 11,000 except in buildups and that he had been advised by Flight Watch that Victor-1 was clear to the east. He said that he was being vectored toward a hole when he encountered severe turbulence in a thunderstorm cell.)

The controller then coordinated a handoff of the Bonanza to another controller. Just after 5:04 he called the Arrow and said, "According to my radar, sir, you're about to enter some pretty severe weather. Uh, do you have it in sight?"

The pilot replied, "Uh, well, we just now getting

some heavy rain now, can we get vectors around this?"

CONTROLLER: Uh, the best thing I can suggest is you, uh, turn right about, uh, 170, and come around to the south and I'll try and get around to the, uh, to the east side of it. It's very solid in your twelve o'clock.

PILOT: Roger, we're getting it now. We're right to 170.

The controller then gave a heading to another aircraft. The next pertinent transmission from the Arrow pilot was his last: "Washington center, 41786 is crashing."

Witnesses reported hearing the aircraft briefly during an electrical storm. High winds, heavy precipitation, and lightning were reported at the time of the crash. Weather radar at Wilmington, North Carolina, depicted a strong thunderstorm over the crash site at the time of the accident.

The horizontal tail and the wings of the Arrow failed in flight. According to the National Transportation Safety Board, all major fracture surfaces were typical of instantaneous overload. This might have been one of the relatively rare cases where an airframe failure in or around a thunderstorm wasn't caused by pilot-induced overloads, as found after a loss of control.

The wreckage was scattered along a bearing of about 075, just slightly to the right of the heading being flown, so the pilot probably hadn't been in the turn long enough to lose lateral control of the airplane.

The weather forecasting system missed this day. Instead of scattered thunderstorms, quite a cluster of activity built up along and near the coast. The sigmets that were issued didn't do the area justice, and the dissemination of information on the situation was weak at best even after sigmets were issued. The sigmets issued were calling for scattered thunderstorms while an air traffic controller was describing a solid line from Wilmington to Fort Mill (near Charlotte, 130 miles inland).

What caused the stronger-than-anticipated development? Instability was certainly there as heat built during the day. The surface temperature at Wilmington reached +34°C before it started cooling down as a result of the thunderstorms. The temperature at 18,000 feet was about −5°C, so the drop of 2 degrees per 1,000 feet for instability was more than satisfied. That rather weak front and the surface low are the keys, though. There was no doubt more lifting than anticipated; it combined with the instability generated by heating to get things going. Once under way, the effects of the outflow from existing cells probably contributed to new development and the construction of quite an area of activity.

As far as air traffic control goes, providing vectors around weather is not a primary duty, and any service depends on the individual controller. In this case one controller thought another wasn't doing the best possible job and moved in to relieve that controller. No person on the ground can *know* what sort of weather situation he is dealing with. The picture might look similar to something that worked a while back, while in reality it has become impassable. This day, some aircraft were being vectored through gaps or holes in the area of weather being presented, but the Bonanza pilot held that he had flown through a cell while being vectored.

The Arrow pilot got the worst of all possible deals. After asking for vectors, he might well have thought that by saying nothing negative the controller had assumed the responsibility of pointing him in a safe direction. But controllers changed and the new one was busy with other aircraft. By the time he got to the Arrow's problem the aircraft was about to enter an area of weather. The warning was not in time, and the Arrow pilot must have penetrated a mature and strong cell. One can theorize that the airframe would not have failed if the airspeed had been at or below maneuvering speed, but the action around a mature cell is such that airspeed control is difficult. A horizontal or angled gust could precede the vertical gust of an updraft and cause a rapid increase in indicated airspeed, leaving the airplane vulnerable to an airframe failure in the updraft.

The simple truth is that it is best to avoid areas of thunderstorms. Trying to penetrate, or to "pick your way" through an area of building activity, has to be considered risky in the extreme—especially when a detour of a hundred or so miles will avoid the whole mess. The fact that it is building is a key, too. That means that new storms will appear along previously clear paths. Holes will close and passages that were acceptable to other aircraft will be inhabited by cumulonimbus shortly afterward.

FH-227 On Approach

It was obvious there was thunderstorm activity in the St. Louis area this July afternoon. The crew of the airline FH-227 approaching the area could see it on their airborne weather radar, and at 5:27 the first officer said to the captain, "We're not going to be able to make it. I don't know, unless we follow it inbound." The captain replied, "Okay," and the first officer asked, "That's okay?" The captain: "Yeah." To which the first officer replied, "It's about 30 miles then from us, between us and the outer marker." The captain added, "About over the outer marker." Clearly there was a problem.

The surface weather chart showed a wave on a quasi-stationary front over northeastern Missouri, and the forecast that was current for the flight made no advance suggestion of a big problem. St. Louis was calling for 3,000 broken, 10,000 broken, good visibility, and south winds at 8 knots, with an occasional ceiling of 3,000 overcast, 6 miles visibility in thunderstorms and moderate rainshowers. There was no sigmet or airmet valid for the area, and the weather radar summary chart issued at 4:40 by the National Severe Storms Forecast Center at Kansas City showed only scattered thunderstorms over the St. Louis area. Earlier that day the NSSFC had determined that the meteorological conditions required for severe thunderstorm activity did not exist.

The automatic terminal information service broadcast that was playing as the flight approached was as benign as the forecast: "Estimated ceiling 4,000 feet broken, visibility 5 miles, haze and smoke, wind 120 degrees at 8 knots, temperature 92, altimeter 30.06, ILS runway 12 right approaches in use, landing and departing runways 12." With this background, the crew's opinion of the weather painted on their radar might not have been as respectful as it could have been.

As the crew flew closer to the St. Louis area, the controller countermanded the ATIS information with word that they would be vectored for an ILS to runway 30 left, which would be more or less a straight-in approach. Between 5:32:26 and 5:39:22 the controller provided vectors through an area of thunderstorms south and southeast of the airport, and at about 5:40 he cleared the flight for an ILS approach to runway 30 left. The flight established contact with the tower shortly afterwards, and there was some discussion about wind and weather. The controller said, "Ozark 809, you're in sight and cleared to land runway 30 Left."

The first officer acknowledged and requested wind information. The controller replied, "Wind is, it's been gusting . . . it's right now, it's 220. It's been around to about 340 degrees, holding at 20 but occasional gusts up to 35."

Shortly, the controller said, "Ozark 809, it looks like a heavy rainshower moving across the approach end of the runway now."

The first officer answered, "Roger, we see it." There was no further discussion about weather.

A TWA Boeing 727 had made this approach about a minute and a half earlier and the captain later stated that because of a strong updraft he had trouble slowing his aircraft to the final approach speed. This captain was unable to establish the desired landing configuration and airspeed, and executed a missed approach. He said he was clear of all clouds about 1,000 feet above the ground and 4 miles southeast of the runway. To the left, the captain of the 727 saw a "wall of water" that paralleled the localizer course and curved around the southwest corner of the airport.

A light twin preceding the 727 landed safely at 5:40. The pilot, though, said that he had difficulty controlling the aircraft after intercepting the localizer. Because of a strong updraft he was, for a while, unable to descend from 6,000 feet. Then a downdraft near the outer marker caused his aircraft to descend at a rate of at least 3,500 feet per minute (the maximum displayed on the vertical speed indicator). He flew out of the downdraft about on the glidepath and several miles from the end of the runway. The NTSB accident report doesn't show that the experiences of the 727 or light twin pilot were passed along to the crew of Ozark 809.

The captain of flight 809 later recalled overshooting the localizer course and disconnecting the autopilot to manually make the corrections to return to course. He could see the runway from a position outside the outer marker and could still see the runway even after the controller told of the heavy rainshower moving across the approach end. As the FH-227 proceeded toward the airport, the captain noticed what appeared to be a roll cloud below, to his left, and parallel to the localizer course. Additionally, there was a "wall of clouds" along the southern and western side of the airport.

The pilot recalled nothing else of the flight except hearing something like hail hitting the airplane, pushing the throttles forward, and applying back pressure to the control column.

An aeronautically qualified witness about 2,000 feet north-northwest of where the FH-227 hit the ground stated that he observed the aircraft making what appeared to be a normal ILS approach. It suddenly climbed 400 to 500 feet and then descended rapidly to 200 feet above the ground. Shortly thereafter, according to the witness, lightning struck the wing just outboard of the left engine. The lightning strike was followed by a rolling flash of fire. The aircraft again lost altitude, and after several apparent "evasive maneuvers" it disappeared into the rain and trees.

A minute before the crash, St. Louis weather reported a special observation: "Measured ceiling 1,100 feet broken, 2,800 feet overcast, visibility 10 miles, thunderstorm, heavy rain showers, wind 300 degrees 29 knots, gusting 30 knots."

Three minutes after the crash it was "measured ceiling 1,100 feet overcast, visibility one mile, thunderstorm, heavy rain showers, wind 220 degrees 20 knots, gusts 33 knots."

The rainfall recorder at the airport showed that heavy precipitation began three minutes before the accident and that 1.55 inches of rain fell in the following 45 minutes. A rainfall recorder located about a mile southeast of the approach end of the runway, and fairly close to the accident site, recorded 1.75 inches of rain between 5:40 and 6:00. Trees surrounding the accident site, 2.3 miles southeast of the airport, were damaged by wind. A National Weather Service expert estimated that surface winds of 65 to 70 miles per hour would have been required to cause the damage. Witnesses 5 miles south-southeast of the accident site reported seeing a mass of debris rotating counterclockwise near the ground between 5:35 and 5:45. The wind in this area damaged trees in an area 450 feet wide and 1,500 feet long and blew the roof off a large building and carried it a distance of about 300 feet.

The National Weather Service at St. Louis did broadcast a severe-thunderstorm warning a minute before the accident, but it was not received by the airline, the crew, or the FAA.

It was clearly a severe-thunderstorm situation, enhanced by the convergence of two distinct squall lines near the airport at about the time of the accident. One line was oriented nearly north–south while the other was aligned east-southeast–west-northwest. The area of activity was moving to the northeast at 30 knots. There was strong instability in the area, and lifting could well have been provided by the low-pressure wave shown on the four o'clock weather chart. The circulation about the wave could have accounted for two separate squall lines. However, from the appearance of the photographs of the weather radarscope at the time, the situation might also have been interpreted as one with severe thunderstorms within clusters or a general area of activity. Whatever the situation, the wave on the front was probably a key.

According to an airline employee who arrived at the scene about 30 minutes after the accident, the captain said that he had been struck by lightning. However, when testifying at the public hearing on the accident, the captain could not recall having made the statement or having been struck by lightning. The NTSB concluded that the aircraft was indeed struck by one or more bolts of lightning, but found no evidence of the lightning strike or strikes adversely affecting any of the vital systems or components of the aircraft.

The activity was apparently more severe along the approach course than at the airport, and there was some suggestion of tornadic activity to the southeast. The switch in surface wind at the airport from 120 degrees (on the ATIS) to 300 degrees to 220 degrees indicated a chaotic situation. Clearly, the airport was affected first by the outflow of one storm and then by the outflow from another. New cell generation was probably continuous, because of the interaction between cells and the general instability in the area. The recording of heavier rain to the southeast of the airport than at the airport also suggests that this airplane got to the wrong place at the worst possible time.

The conditions encountered by the FH-227 (and the preceding airplanes) are a good outline of what happens along the leading edge of an approaching severe storm. Updraft activity affected all three airplanes. The light twin encountered a downdraft and probably windshear but flew out of it before getting too close to the ground. The 727 apparently experienced only the updraft but got away in a missed approach. As for the FH-227, the NTSB suggested it was possible that the captain initially descended below the glidepath in order to remain below the clouds and to maintain visual contact with the ground during the approach. In any case, the final descent into the ground was most likely the result of an encounter with the dynamics of a severe thunderstorm.

The wind shift from northwest to southwest could have been a major factor. The northwest wind was a direct headwind, the southwest wind a crosswind. If the wind at the aircraft's altitude also made an abrupt shift from northwest to southwest, the result would be a decay in airspeed (the result of a rapidly decreasing headwind component). Throw in the downdraft and it is easy to imagine the final sink from which the crew was unable to recover. This is also a good example of turbulence well clear of the heaviest precipitation. The 727 was apparently well clear of precip, there was no mention of it by the light twin pilot, and the FH-227 captain said that he could see the runway from a position outside the outer marker.

WITHOUT WARNING

The instrument-rated pilot checked conditions for a flight from Huntsville, Alabama, to Daytona Beach, Florida, by calling the National Weather Service office in Huntsville at about eleven o'clock on a July morning. The briefer informed him of IFR conditions at Huntsville that were expected to improve to VFR within two hours. The latest available radar summary chart showed some thunderstorm activity on the west coast of Florida, but nothing along the route to Daytona. The briefer added that scattered thundershowers were expected to develop in southern Georgia and Florida in the afternoon and that the Daytona forecast called for a chance of thunderstorms.

So informed, the pilot called the Muscle Shoals Flight Service Station and filed IFR for the flight in his Bonanza. He requested a cruising altitude of 7,000 feet, and shortly before 1:00 P.M. the Bonanza was at that altitude, cruising routinely toward its destination.

Radio communications with the Bonanza were related mostly to frequency changes. Once, the controller did ask the pilot to pass a message along to another airplane, and there was a short radar vector to get the airplane around some other traffic. At about 1:30 the Bonanza was cleared direct to the Eufaula (Alabama) Vortac. Eighteen minutes later the controller in the Atlanta center handling the flight talked to the Jacksonville center controller working the next sector through which the Bonanza would pass.

ATLANTA: Bonanza 7623 November is three miles southeast of Eufaula 7,000.

JACKSONVILLE: Seven southeast.

ATLANTA: Seven southeast.

JACKSONVILLE: Ha, let's check another one here.

ATLANTA: Do you see him?

JACKSONVILLE: Just terminate him. I don't see him.

ATLANTA: Bonanza 7623 November, radar service terminated. Contact Jacksonville center 128.1.

BONANZA: 7623 November, 128.1.

ATLANTA: Roger.

The Bonanza pilot did not call Jacksonville as instructed. A few minutes later, the controller tried to establish contact with the aircraft, to no avail. A Flight Service Station attempted contact, and personnel at the approach control facility at the Albany, Georgia, Naval Air Station (southeast of Eufaula) were asked if they saw the aircraft on radar. It was not until 2:10 that anyone mentioned weather. Then the Jacksonville center controller said on an interphone line to the Jacksonville center flow control position, "I talked to Navy Albany and they looked out there on radar between Albany and Sawmill [an intersection on the airway]. They said they can't see anything, even a primary, and we have a lot of weather in there, so I got radio on him and he's overdue at Albany and radio hasn't been able to get him either."

At 2:50 the controllers got an accident notice on the airplane. It had broken up in flight and crashed about eight miles southeast of Eufaula, just south of the airway along which it had been cleared. Time of the accident was only moments after the last contact with the airplane. The in-flight structural failures were indicative of pilot-induced overloads. Wreckage distribution suggested that the in-flight breakup occurred at a relatively low altitude compared to the 7,000-foot cruising level. The aircraft was probably on a southerly heading at the time of the breakup.

Witnesses in the area reported that there was thunder and lightning and that they heard a loud roar before the airplane hit the ground.

The 1:30 radar report from the National Weather Service station at Waycross, Georgia, showed 40 percent coverage of strong echoes containing thunderstorms producing heavy rainshowers in the accident area. The maximum top was 45,000 feet; this was approximately in the location where the Bonanza was lost. The weather radar at Apalachicola, Florida, showed 20 percent coverage in the area with a max top of 43,000—also in the approximate location of the crash. A military airplane reported ten minutes after the accident that it was "bad to the southeast of Maxwell [which is northwest of Eufaula], need radar to go 20 miles northeast of Tuskegee, line 200 miles long, tops 15,000–20,000, building."

The optimistic forecasts given the pilot were clearly in error, and no further information was offered him, even though the air traffic control radar was apparently showing activity in the area.

Why did the thunderstorms develop so rapidly in an area where there had been none, and not a lot of activity was forecast?

There was a simple and obvious reason for the activity. A low aloft, at the 500-millibar level just southwest of the area, resulted in instability and good support for thunderstorms, and for the intensification of thunderstorms as the day wore on and the surface heated. A weather chart prepared after the accident showed a surface low-pressure trough in the area, too, and this could well have been developing at the time the Bonanza pilot tried to fly through the area.

The general weather in the area from which the pilot flew was excellent. Columbus, Georgia, just north of Eufaula, reported 3,000 scattered, 25,000 scattered, and 7 miles visibility. So as the pilot flew toward Eufaula, it is likely that he was clear of clouds with a reasonable view of things ahead. That couldn't last, though, because both Dothan (south of the accident site) and Albany (southeast) had much more congested cloud conditions. Albany had a broken layer at 2,500 and Dothan also a broken layer of approximately that height. Both stations also had higher clouds. Dothan was reporting thunderstorm activity to the north and northeast, and Albany had rainshowers of unknown intensity northwest through north. Given these conditions, it is fairly sure that the pilot was flying into an area where he might not be able to see and avoid cumulus or cumulonimbus.

If the fact that weather was never mentioned by the pilot to a controller is given full weight, perhaps this pilot was in no way apprehensive about what lay ahead. It's hard to imagine, though, that with a heavy thunderstorm topping out at from 43,000 to 45,000 feet ahead, the sky wouldn't have had an ominous appearance from 30 to 40 miles away. Could the pilot have thought that no mention of weather from the air traffic controllers meant there was no weather in the area? The pilot had been flying for a number of years and had almost 1,000 hours. His instrument experience wasn't broad, though, and he wasn't a resident of the southeast, so he probably did not appreciate the peculiarities of local weather in the area.

Regardless of what led the pilot to continue, it is fairly clear that he got into or very close to a thunderstorm cell. Ground witnesses plus the Waycross and Apalachicola weather radar reports all agree that something was going on in the area. The altitude and nature of the inflight breakup suggests that the pilot could have entered the area of weather and then started a right turn, which would have been toward better weather. Control of the airplane was apparently lost in this right turn and the airframe failed at low altitude, possibly when the pilot recognized a high-speed (in excess of the redline) diving spiral and attempted a recovery. There was no evidence that the pilot extended the landing gear to minimize speed buildup.

This day, there was available weather information (the 500-millibar chart with the low depicted at that level) to strongly suggest an outbreak of thunderstorm activity in the area where this airplane was lost. The forecasts didn't pick this up, and when weather did develop, the air traffic controllers volunteered no information. That put the burden on the pilot. It was a clear illustration of the meaning of the federal aviation regulation that "the pilot in command of an aircraft is directly responsible for, and is the final authority as to, the operation of that aircraft."

VFR TWIN

The instrument-rated pilot of a 310 filed a VFR flight plan from Grand Rapids, Michigan, to Grand Island, Nebraska, about an hour after being briefed on below-VFR conditions in fog with areas of scattered thunderstorms in Iowa and Nebraska. By the time he filed the flight plan, the visibility had improved to marginal VFR in haze, according to the flight service specialist. The month was August. Takeoff was at 10:02 A.M. The surface weather chart showed a cold front from northern Kentucky to southern Illinois with a trough of low pressure from the Iowa-Missouri border into extreme southeastern Nebraska.

The weather was marginal VFR at best as far as Cedar Rapids; past that point, conditions deteriorated. Des Moines, just south of the aircraft's route of flight, was reporting 1,300 broken and 12,000 overcast with 2½ miles visibility. Omaha, a little farther along, was reporting IFR conditions all morning. Sioux City, north of course, held on to very marginal VFR conditions until noon, when it went IFR.

Because this pilot was flying VFR, there's no way of knowing what his method of operation had been until he called the Omaha Flight Service Station at 12:45 and reported 32 miles east of the Neola VOR. The pilot reported flying at 10,800, VFR on top, westbound, with higher buildups ahead. He requested the Sioux City weather, to divert north around buildups or to Sioux City. Sioux City was, at the time, reporting 2,500 broken, 4,000 overcast, 4 miles visibility in light rain and fog. The Norfolk, Nebraska, weather was also furnished to the pilot. It was 500 scattered, 4,000 overcast, 7,000 overcast with 7 miles in light rain. A pilot report on between-layers conditions at 9,000 feet north of Norfolk was furnished, as was a radar report indicating activity along the pilot's route of flight and a Chicago airmet dealing mostly with low ceilings and IFR conditions. The FSS specialist suggested that the pilot contact Sioux City approach control for possible radar assistance around the thunderstorm buildups. The pilot acknowledged this information and said, "Will go north around Sioux City to VFR conditions west of Sioux City and Norfolk."

At 12:51 the pilot contacted Omaha FSS again and requested the Grand Island weather. It was 400 scattered, 900 broken, 1,600 overcast with 4 miles visibility. The specialist volunteered the latest Sioux City weather as well: 900 broken, 1,500 overcast, 1½ miles visibility in light rain and fog. The pilot acknowledged this information, and Omaha FSS had no further contact with him.

At 1:27 the pilot attempted contact with Sioux City approach control. They heard one call and tried to answer. Then at 1:31 the pilot transmitted: "Sioux City Approach, 62 Tango, ah, ah, Mayday please."

Sioux City acknowledged and read off the current weather but the pilot did not answer. Then approach control called a Minneapolis center controller with radar coverage in the area and said, "We got an airplane that called us a while ago with a Mayday and we can't establish contact with him. Do you see that target, ah, oh, ten to fifteen miles southeast of Rodney circling?"

Minneapolis replied, "I got one 15 southeast of Rodney but it's not circling."

There was discussion about a target in the area, with no agreement on whether they were both looking at the same one and that it might be the aircraft that had called Mayday. In a discussion about whether or not there was a primary target (as opposed to a transponder return), Minneapolis center said, "Well, with the weather out there it is hard to see." Sioux City agreed that the weather in the area made it difficult to see returns from aircraft on radar.

The assistant chief at Sioux City heard the Mayday

call and went to an unused radarscope in an attempt to locate any unidentified aircraft that might have been squawking 7700. The only target he observed was a beacon-equipped aircraft 40 southeast of Sioux City, moving toward the northwest. He watched the target move to a point approximately 35 southeast, then make a 180, fly southeast-bound for 5 to 7 miles, and then disappear from the scope.

Weather radar data showed considerable thunderstorm activity in the area. It was described as "very strong" by the radar meteorologist in Des Moines, with maximum tops as high as 40,000 to 42,000 feet.

Two witnesses said they heard the sound of an airplane flying low in the area and that it would sound normal and would then sound like an engine "pulling down." Another witness reported a heavy rainstorm with lightning. Later his wife looked out the window and saw something in their field. Investigation revealed that it was the tail of the 310, which had crashed in a right-wing low, level attitude in the cornfield and had come to rest in a tree line on a southerly heading. The landing gear was down at the time of the accident.

When flying toward an area of thunderstorm activity, especially one connected with a front, the normal progression would be to encounter cumulus with ever higher tops. The closer to the front, the more lifting action. Given instability of the air, the development would become more pronounced with more lifting. In this case, it was a trough instead of a front, but a trough can provide lifting adequate to trigger thunderstorms where the air is unstable.

This pilot's reaction to the increasing tops was to divert northward. The weather that way wasn't exceptional, though, with layered clouds between the reported 10,800-foot cruising level of the airplane and the ground.

That the pilot called Sioux City might have been an indication that he intended to land there. But contact was never established. Nor was a definite correlation made between the Mayday call and the disappearance of the unidentified target from the Sioux City radarscope.

If indeed that target was the 310, the pilot was apparently dissuaded from continuing directly to Sioux City by weather between him and the airport. The 180 was successful. There's no way to determine with certainty that the airplane was actually in or near a thunderstorm cell at the time it crashed, but it seems highly likely given the radar observations of the time. The pilot had extended the landing gear of the airplane, which is a good thing to do when turbulence becomes severe and control doubtful. If control of the airplane is lost with the gear down, the speed will build more slowly, allowing more time for recognition and correction of the situation.

There is no apparent reason why the pilot didn't revert to IFR. Certainly there is a school of thought that thunderstorms are best handled VFR and down low, but the prerequisites for this include reasonably good visibility outside the storms, and an airplane that is flying at low level to begin with. Neither was the case. The weather was cruddy and the pilot started out above 10,000 feet. One can only assume that the pilot had descended to a lower altitude. The storms this day were probably not severe, but they were embedded in other clouds and in general rain. It would have been a very difficult time for picking one's way through VFR.

Violence Clear Of The Storm

The experienced pilot checked weather at 5:30 P.M. for a flight in a Navajo from Trenton to Indianapolis. There was a low far to the northwest of Indianapolis with an occluded front pushing into Indiana at about 20 knots. Some thunderstorm activity was called for in the area forecast, but that document ended with a hopeful "Gradual clearing near sunset." The terminal forecast for the Navajo's Indianapolis ETA was for 2,000 broken, wind 240 at 10 knots, broken clouds variable to scattered. The only sigmets or airmets current for the route at the time of briefing covered low ceilings in light rain over western Ohio and eastern Indiana. These conditions were moving eastward. There was no mention of thunderstorms.

The flight departed Trenton at 6:23 P.M. Departure was VFR, but at 7:19 the pilot called Harrisburg Flight Service Station (about 100 nautical miles west of Trenton) and said he "thought he had better file IFR, and that he was 30 miles west of Harrisburg." Transmissions from the aircraft were weak and the FSS specialist had to have the pilot repeat a number of the items. Enough information was received to enter a flight plan in the computer, and the pilot was told to contact Cleveland center.

The clearance to Indianapolis was routine, as was the initial portion of the flight. The pilot reported being in some light rain for a while, and later, as he progressed westward, he mentioned being on top. Farther west, near Columbus, Ohio (about 170 nm from Indianapolis), the pilot mentioned seeing a lot of lightning in the distance. The Columbus controller said he was painting nothing on radar.

The lightning the pilot saw was, in fact, near Indianapolis. And the terminal forecast the pilot had received was fast becoming inaccurate as an area of severe activity bore down on Indianapolis. Conditions were, in fact, terrible. As the Navajo neared Dayton (about 90 nm from Indianapolis) the pilot of a Piper Seneca was learning just how wild things can get in the vicinity of a severe thunderstorm.

The Seneca pilot was inbound to Indianapolis. While being vectored south of a weather area north of Indianapolis and while descending to 3,000 feet, this pilot encountered what he later called a "situation unique . . . in all my time of flying. I have flown many hours with severe turbulence such as mountain flying in California during thunderstorms but nothing quite as unique as what I encountered south of Indianapolis airport on Sunday evening."

The pilot had, fortunately, slowed the Seneca to 130 knots before encountering turbulence so severe that he reported it to approach control as "treacherous. It was as if I had been thrown into a tornado."

This pilot wrote, for the NTSB record: "Immediately the airplane attempted to go into a spin. I took it off autopilot in order to maintain control. The plane yawed 60 degrees to the north, losing lift and trying again to go into a spin. It then yawed back 180 degrees to the south, again losing lift and going into a spin. It was all I could do to maintain the airplane in an upright position and impossible to control the direction of the aircraft. I again notified approach control that it was terribly severe turbulence. I was then cleared to descend to a lower altitude, and on reaching 2,000 feet the aircraft seemed to become a little more controllable, even though it still seemed as if it were attempting to go into spins. At least six or seven times I found myself in a partial spin, but finally at about 1,400 feet I was able to level out and get directional control."

The pilot added, "Oddly enough, the skies were clear with lightning and clouds visible to the north. Being in such a clear area one would have no warning to expect such severe turbulence."

The radar at Indianapolis center was inoperative this evening, and as the pilot of the Navajo was passing Dayton he had an extensive discussion with the Dayton Flight Service Station about the weather ahead.

Dayton said, "We are looking at our radar scan here, which shows an intense thunderstorm, which is just west extending—north of Indianapolis. It's moving eastward about 30 knots. Looks like Indianapolis is going to get the southern edge of it here and it's probably the lightning you see. It looks like it extends from west of Muncie, westbound, and it runs north of Indianapolis for, oh, looks like about 75 miles.

"Indianapolis right now 4,000 scattered, visibility 15 miles, and they're reporting this buildup northwest through northeast and it's moving east, with occasional lightning in clouds and cloud to ground. I got a pilot report just west of Indianapolis—he's picking up severe turbulence at 3,000 feet and Indianapolis has amended their forecast between now and 2300 (EST) for 2,000 overcast, five in light showers, a chance of 400 foot overcast, a half a mile, thunderstorms, which will be heavy, with hail, expected gusty winds up to 40 knots with the thunderstorm activity, and it looks like it's going to be in Indianapolis—right in the vicinity of Indianapolis in, oh, half an hour."

The pilot replied, "Okay, we'll, I'll have to play that a little bit, won't I."

The FSS specialist said, "Affirmative, now, they have been, this cell been watched across Illinois, move into Indiana, and they were ahead. Have been reports of hail with this particular system, and also, ah, there was a tornado, ah, reported with it . . ."

There was further discussion, with the pilot indicating that maybe he could go around the weather. There was also some conversation about Indy center's radar being out.

After the pilot of the Navajo switched from flight service to an air traffic control frequency, the Dayton FSS rang the control tower at Muncie, Indiana (51 miles northeast of Indianapolis), and asked about the weather there. The person answering at Muncie tower said the weather appeared to be about 10 miles from the airport and added, "It looks like it's going to be heavy stuff." Tornado sightings were discussed, and the tower controller said they were ready to abandon the structure.

At 9:24 the Navajo checked in with Dayton approach control, and discussions about weather continued.

Some airline aircraft were on the frequency and were also involved in the discussion of weather.

In reply to a question about the weather looking good on into Indianapolis, the Dayton controller told the Navajo pilot, "Ah, by the time you get there, it won't. I suggest deviate about 10 miles south of course. There's a line moving from Muncie that's extending southward now, just north edge of Victor-50 right now."

The pilot suggested that he might best go directly to Shelbyville, which is southeast of Indianapolis and south of Victor-50. The controller approved this.

The pilot, perhaps seeking some assurance, then said, "Alright, and that'll get me, thank you very much, and that'll keep me out of it then, won't it?" The controller only agreed that it would keep the flight out of "most of it."

Next came some conversation between the controller, a TWA flight, and an Allegheny (now USAir) flight. The TWA was inbound to Indianapolis and the Allegheny had just departed from that airport. The Navajo pilot was also on this frequency.

> TWA: Are you showing anything between us and the field?
>
> CONTROLLER: I gotta pretty heavy cell right over Indianapolis VOR, which is about 10 miles north of the airport. You should beat it in, though.
>
> ALLEGHENY: Aircraft just calling, is he going into Indy?
>
> CONTROLLER: Ah, I've got one just west of Shelbyville going to Indianapolis, yes, sir.
>
> ALLEGHENY: Okay, he's got, ah, this is Allegheny-

ny, we just came out of Indy and, ah, coming out eastbound directly from Indy we had quite a good windshear and, I'd say, severe turbulence, so his best bet is to sneak in on the south side of it and my advice if he gets turbulence to turn around and land someplace else.

TWA: Oh, it was that violent, huh?

ALLEGHENY: Well, it was severe.

TWA: That sounds good enough to me. I think we'll go on down to Cincinnati.

The captain of the Allegheny flight later gave the NTSB the following statement about the turbulence: "While climbing out of approximately 7,000 feet, we encountered severe turbulence. It was only with great difficulty that we were able to maintain a reasonable flight attitude. In my opinion, I feel that the situation would have been in grave doubt had I not been flying a heavy airplane.

"There were scattered-to-broken clouds in our area, with no cells painted on our radar. I requested, and was granted, an immediate turn to the south, to increase our distance from the weather to the north. We were out of the turbulence in approximately one minute and were able to turn on course without further incident. I reported the weather condition to the controller."

From Dayton, control of the Navajo was transferred to the Indianapolis center. Its radar wasn't working well and about the only solid information the center could give the Navajo pilot was that the TWA 707 had diverted to Cincinnati rather than go to Indianapolis. The pilot might not have heard the previous discussion between TWA and Allegheny, or at least might not have followed the conversation, because when told of the diverting 707 the pilot said: "Alright, thank you very much. That's good news."

At 9:40, control of the flight was transferred to Indianapolis approach control. After the initial contact, the conversation went as follows:

CONTROLLER: Nine Mike Charlie, you are in radar contact, 22 miles east of Shelbyville VOR.

NAVAJO: Roger, and, eh, do I look on in . . . if it's . . . if it's VFR in there I might land on 31 to avoid, or to get in quicker, four, eh, are they running four, you say?

CONTROLLER: Nine Mike Charlie, at the present time most of the weather I am showing is north of the Shelbyville VOR at, eh . . . is moving east southeastbound at the present time.

NAVAJO: I'm looking . . . a few clouds here now . . . but it's not too rough yet.

Moments later, from the Navajo: "Indianapolis . . . if you let me down, how soon you going to let me down to maybe three or four?"

CONTROLLER: Nine Mike Charlie, your transmitter is garbled on this frequency, come up on 121.1.

NAVAJO: Roger. . . . Forty-nine Mike Charlie with you at 6,000.

CONTROLLER: Four Nine Mike Charlie, maintain 6,000 and you copy a lot better on this receiver. . . . Navajo Nine Mike, descend and maintain 5,000.

NAVAJO: Roger, thank you, sir. . . . Four nine

Mike Charlie, I hit a ton of weather out here. My heading is 210 degrees.

CONTROLLER: Four nine Mike Charlie, roger.

NAVAJO: Will I be through this stuff in a minute?

CONTROLLER: I'm showing you south of all the heavy weather at this time.

NAVAJO: I'm south of all the heavy weather? Boy this is rough here.

That was the last transmission from the Navajo. Shortly after, it crashed in an open field, in a near vertical attitude.

A farmer who lived about half a mile from the accident site described the situation: "Sky was clear, could see stars and moon. The wind started gusting to 25 to 30 miles per hour at the time I heard the aircraft. The engines sounded as if they were pulling a heavy load. It sounded as if they were heading northwest. At that time I heard the engines miss out, and at that time I ran out the back door.

"I forgot to say that there were thunderstorms to the north and northwest, about 20 miles off. Just before we heard the aircraft there were low black clouds moving to the southeast.

"I looked to the north-northwest, at which time I saw the aircraft start to a slow pitch downward. Then it went straight nose downward to the ground."

Though they didn't see the aircraft, a couple of pilots traveling in automobiles gave graphic descriptions of weather conditions in the area.

One said, "I observed a front rolling in from the

north-northwest, very fast. . . . Light rain fell for three or four minutes, with strong wind, followed by a calm period for one or two minutes. Rain, light hail, and strong wind then began."

The other driving pilot's statement included the following: "Just to the northwest of Rushville was a line of cumulus clouds running northeast to southwest, tops approximately 15,000 feet. There was no lightning in these clouds; however, when I drove under the line . . . there was light rainshowers and very ragged bottoms, approximately 3,000 feet or higher. The clouds had that strange look commonly associated with turbulent conditions."

Even though the tops of the storms were not exceptionally high, and even though the National Weather Service never actually issued a sigmet, the situation was clearly severe and there was plenty of information available on the severity of the weather.

The air was unstable. The low to the west was reasonably strong, and there was a closed low at the 500-millibar level that slowed the eastward progress of the surface low and had earlier allowed it to intensify. The surface low moved only from one side of Minnesota to the other in 24 hours (12 either side of the accident time), but the circulation around it was strong. It was a very dynamic, fast-moving situation that resulted in an occluded front—cold overtaking and pushing under warm. This front moved rapidly eastward, and 24 hours later it was oriented east–west out of the surface low in southern Minnesota. The low was filling by this time, and both the front and the low had disappeared from the following day's map even though the closed low aloft remained.

The winds aloft at the time of the accident were westerly to southwesterly up to 14,000 feet, and the area of activity the Navajo was avoiding was moving from 280 degrees at 30 knots. The classic severe-storm scenario calls for strong southwesterly winds aloft and a low-level southerly flow, but there can certainly be exceptions. Perhaps it was the lack of this classic pattern that kept the forecasters from indicating thunderstorm problems earlier, and that resulted in no sigmet being issued and no severe weather warnings going to aviation interests. They must have missed the significance of the closed low aloft to the west and the occlusion.

As the Navajo traveled south of the weather, it passed about 16 nautical miles away from a heavy cell about four or five minutes before the accident. At the time of the crash the airplane was about 20 south of a couple of moderate cells, as painted on weather radar at Cincinnati. The witness reports suggest roll cloud activity in the area of the accident, and apparently there was extremely strong inflow and outflow connected with the thunderstorm activity. The Allegheny flight encountered severe turbulence at 7,000 feet, the Seneca on a descent to 3,000 feet, and the Navajo at 5,000 to 6,000. (The elevation in the area is about 1,000 feet; subtract this for altitudes above ground level.)

The Navajo didn't suffer any in-flight damage, according to the NTSB report. It just dove almost vertically into the ground. From the Seneca and Allegheny pilots' descriptions of conditions in the area, the shear turbulence was extreme. The pilot could have lost control in the turbulence or could have stalled the airplane after slowing it in an attempt to improve the ride.

The significant lesson is in the distance of the three aircraft from precipitation when two had trouble and one crashed. They were not that close to the cells and were nearly if not totally in compliance with the guideline to stay 20 miles away from cells when severe weather is forecast. This was indeed a time when conservative intuition, such as that displayed by the TWA crew that diverted to Cincinnati, was the saving grace. Radar and guidelines simply were not adequate. The appearance of the low clouds ahead of the storm and visible in moonlight might have been a clue, but it is tough to form a really good visual picture with nothing more than moonlight.

The time of frontal occlusion can be violent, and perhaps the Navajo reached this area as the cold front was just beginning to push under the warm front, resulting in severe turbulence a good distance ahead of observed thunderstorm activity.

The Severe Storm And The DC-9

On April 4, 1977, Southern Airways (since merged into Republic) lost a DC-9 in a classic encounter with a severe thunderstorm. The accident was especially dramatic because of the completeness of the cockpit voice recorder tapes, and because both the DC-9's engines failed in the thunderstorm and the crew had the opportunity to land the airplane after the encounter. There were survivors, so the landing has to be judged at least partly successful—especially when you consider the ramifications of a power-off landing in an airplane like a DC-9. But it was the weather associated with the accident and the aircraft operations to the point of thunderstorm penetration that hold the weather lesson, so I will concentrate on that aspect of the event.

To begin with, I had some knowledge of the severity of this system from a flight through it the day before, in a Bonanza. The activity was farther west that day, and on a flight from Mobile, Alabama, to Wichita and then back to Little Rock I got a double dose.

First came a passage through the front after detouring west of a straight line from Mobile to Wichita. At this time, the bases of the thunderstorms seemed high and I had a relatively smooth trip through at low altitude. There were only a few minutes of turbulence. I used a combination of the traffic controller's radar and an airline pilot's reports in picking my path. The airliner was operating at a very low altitude (5,000 feet as I remember) because of the weather. The pilot had obviously been working the area during the day on his short-haul flights, and had found this the way to go.

On the other side of the activity, northwest of Little Rock and on toward Wichita, conditions were much better. But I could see the effects of the jet stream west of the weather. High clouds were moving *very* fast, and looking to the west it appeared that I could see the tip of the trough aloft, where high clouds turned from southeastbound to eastbound and then to northeastbound.

That sighting could well have been an illusion, but I still remember it well.

Coming from Wichita back to Little Rock, I found that the thunderstorm activity had actually moved back to the northwest. It had been southeast of Little Rock earlier; I had to fly through 20 or more miles of it getting back into Little Rock from Wichita later in the day. The strong jet stream activity developing to the west of what had been a reasonably benign front could have had the effect of initially pulling surface weather back toward the west. It clearly had all the signs of something big brewing.

The crewmembers of the Southern flight spent the night of April 3 in a motel in Muscle Shoals, Alabama. Their first flight of the day was from Muscle Shoals to Atlanta, where they arrived at 9:20. They departed Atlanta at 10:51 A.M. on a series of flights that led them back to Muscle Shoals for the ill-fated 3:21 departure to Huntsville, Alabama, and Atlanta. The crew had just two hours previously flown the route they were about to fly and, according to the NTSB, they probably relied more on their knowledge of actual conditions than on a forecast or warning of conditions that might materialize.

At Muscle Shoals they received from Southern's flight dispatch system two tornado watches and two sigmets. These covered forecasts of conditions that were expected to materialize in northern Alabama and Georgia between 1120 and 2000. The sigmets called for scattered to numerous thunderstorms, occasionally in lines, a few severe, possibly a tornado, occasional tops to 45, 000 feet. The tornado warnings covered the route of flight to Atlanta and called for possible tops to 58,000 feet. A line of thunderstorms from southwest Mississippi to northern Alabama was to continue to intensify while a small-scale low center moved northeastward from southern Mississippi. This was all typical of what might happen along the east flank of a trough aloft, with strong jet stream activity.

The stop in Huntsville was routine. The crew remained in the cockpit of the DC-9 and the only additional weather information given them was the current observation for selected terminals.

According to the NTSB, when the flight departed Huntsville, the crewmembers had little meaningful weather information to alter their impressions of conditions that existed when they flew the route two hours earlier. They had weather warnings, but nothing concrete that told them there were actually storms along the path. The NTSB noted that despite regulations requiring pilots and dispatchers to get and use all available information there was "no evidence that either the flightcrew or flight dispatch personnel made any significant attempt to seek information on current conditions along flight 242's route between Huntsville and Atlanta, including information from the 2:59 P.M. report from Rome which identified thunderstorms to the northeast and southwest of Rome."

The flight departed Huntsville at 3:54 P.M., to fly direct to Rome VOR and then on into Atlanta. Huntsville had reported a thunderstorm overhead moving northeast at about the time of departure. A few minutes later Rome was reporting continuous thunder southwest through northwest, with the pressure falling rapidly.

Weather radar had, about 20 minutes before the flight departed, recorded an area of very strong radar echoes in the area with the maximum tops at 46,000 feet about 35 miles northwest of the Rome VOR. Again, the crew did not have the Rome weather or this radar report.

The progress of the flight toward this weather is best tracked through following pertinent weather-related conversation from the cockpit voice recorder.

CAPTAIN TO FIRST OFFICER: Well, the radar is full of it, take your pick.

HUNTSVILLE DEPARTURE CONTROL TO 242: Southern 242, I'm painting a line of weather which appears to be moderate to, uh, possibly heavy precipitation starting about, uh, five miles ahead and it's . . .

242: Okay, uh, we're in rain right now, uh, it doesn't look much heavier than what we're in, does it?

HUNTSVILLE: Uh, it's painting . . . I got weather-cutting devices on which is cutting out the, uh, precip that you're in now, this, uh, showing up on radar, however it doesn't, it's not a solid mass, it, uh, appears to be a little bit heavier than what you're in now.

FIRST OFFICER TO CAPTAIN: I can't read that, it just looks like rain. Bill, what do you think? There's a hole.

CAPTAIN: There's a hole right here. . . . That's all I see. . . . Then coming over we had pretty good radar. . . . I believe right straight ahead, uh, there the next few miles is about the best way to go.

Moments later, the captain suggested hand-flying the airplane if it got rough.

HUNTSVILLE: Southern 242, you're in what appears to be about the heaviest part of it now. What are your flight conditions?

FLIGHT 242: Uh, we're getting a little light turbulence now and, uh, I'd say moderate rain.

HUNTSVILLE: Okay, and uh, what I'm painting, it won't get any worse than that and, uh, contact Memphis center on 120.8.

Southern 242 switched over to Memphis and checked in on a routine basis. At 3:58:22 the captain said to the first officer, "As long as it doesn't get any heavier we'll be all right." Moments later the center made a broadcast advertising the existence of a new sigmet covering hazardous weather in the area. This was greeted by an in-cockpit "Oh, shucks" epithet from the captain.

The control of the flight was then handed off to Atlanta. The check-in at 3:59:06 P.M. was routine, but the captain had just said, "Here we go, hold 'em cowboy," to the first officer.

At 4:01, the National Weather Service station at Athens, Georgia, made a special radar report of intense echoes containing thunderstorms with intense rainshowers. The center of one group of cells was 15 nautical miles west of the Rome VOR. The maximum top was 51,000 feet, and the group of cells was 10 nautical miles in diameter, and the cells were moving east-northeastward at 55 knots. Flight 242 was approaching Rome from the west-northwest and had no information on this activity or on the fact that there had been a tornado outbreak in Alabama over the past couple of hours and that a tornado had passed through the southern suburbs of Rome at about 4:00 P.M. In fact, looking back after this eventful day, the National Weather Service determined that this storm system was one of the most severe in three years, one of the fastest-moving on record, and that the 30 severe thunderstorms associated with it spawned 20 tornadoes.

Southern 242 was flying at 17,000 feet when a TWA flight had the following exchange with the controller:

TWA: Uh, center, TWA's 584, this, this is really not too good a corridor we're coming through here. It's too narrow. . . . Uh, we're getting moderate, uh, heavy moderate turbulence and quite a bit of precip in here.

CONTROLLER: 584 roger, it looks like, uh, right now another 15 miles to the south, you should be through the, uh, southeastern edge of what I'm showing and, uh, maybe a little better.

TWA: Okay, it's good to have hope anyway.

There was further discussion between the controller and TWA 584 about the weather. At 4:02:57 the Southern pilot said to the First Officer, "I think we had better slow it up right here in this, uh . . ."

The First Officer answered, "Got ya covered." At this time the aircraft was nearing the area of intense activity.

The conversation was all about weather starting at 4:03:48, when the airplane was about 20 nm from an extreme radar return.

CAPTAIN: Looks heavy, nothing going through that. . . . See that.

FIRST OFFICER: That's a hole, isn't it?

CAPTAIN: It's now showing a hole, see it.

The sound of rain on the windshield was recorded at 4:04:05, as the airplane entered an area of light rain that was being painted on ground weather radar. At this time the airplane was pointed directly at the heaviest rainfall in the system, about 15 miles ahead.

FIRST OFFICER: Do you want to go around that right now?

CAPTAIN: Hand-fly at about 385 knots.

(The first officer was apparently handling the controls of the DC-9 throughout.)

At 4:04:30 the sound of hail and rain was recorded, and shortly thereafter the captain informed the controller that they were slowing up a little bit.

At 4:05:53 the airplane was about 7 miles from the core of the storm and still headed toward it.

FIRST OFFICER: Which way do we go, cross here or go out? I don't know how we get through there, Bill.

CAPTAIN: I know you're just gonna have to go out . . ..

FIRST OFFICER: Yeah, right across that band.

At 4:06:01 the area of strongest return was slightly to the left of the nose of the DC-9, with lighter return to the south.

CAPTAIN: All clear left approximately right now. I think we can cut across there now.

FIRST OFFICER: All right, here we go. . . . We're picking up some ice, Bill.

At 4:06:42 the controller cleared Southern 242 to descend to 14,000 feet. Fifteen seconds later the sound of hail and heavy rain was recorded and was the only sound recorded by the cockpit microphone until the power to the recorder was interrupted at 4:07:57. The power was off for 36 seconds; when it was restored the sound of rain continued for 40 seconds. At 4:08:37, the first officer said, "Got it back Bill, got it back, got it back."

At 4:09:15 the captain called the controller and said, "Okay, uh, 242, uh, we just got our windshield busted and, uh, we'll try to get it back up to 15, we're 14." Within a minute they informed the controller that they had lost both engines and requested a vector to a clear area. The rest of the conversation related to the forced landing.

According to one of the surviving passengers, a pilot with a commercial license, the flight was routine until the aircraft encountered severe turbulence followed by very heavy precipitation, a lightning strike on the left wingtip, and hail. The hail increased in intensity and size; then the right engine quit, followed by the left engine. He estimated that the turbulence lasted from one to two minutes, the heavy precipitation lasted from 45 to 60 seconds, and the hail lasted from 45 to 60 seconds.

That the airplane penetrated the area of severe weather and emerged from the other side still intact and in control is tribute to the strength of the airframe and the skill of the first officer at hand-flying a DC-9 in severe turbulence. Why, though, did the flight penetrate the area of severe weather?

First, one must consider that all flightcrews do not

follow the scriptures about giving cells a 20-mile berth whenever severe thunderstorms are forecast. Southern 242 was not the only airline aircraft operating in the area; it just happened to be the only one that penetrated the worst part of the storm at the worst possible time. But for the grace of God, any of the other aircraft in the area could have done the same thing.

In the transcript, the captain suggested a cut to the left at just the wrong time. The slight turn took the aircraft toward worse instead of better conditions. He certainly didn't do this on purpose, and there is a logical reason why a pilot with perfectly good airborne weather radar might have done this.

Attenuation (the absorption and/or scattering of the radar beam by precipitation) was addressed in the manual for the radar installed in this aircraft as follows: "Severe rainfall within the antenna near field (100 feet) disperses the beam with a consequent reduction in range performance. Radome icing reduces system range performance. In severe cases, all targets disappear, and an indistinct haze may appear at the indicator origin." The airplane was flying in rain before reaching the area of most severe activity, and it was picking up airframe icing.

Also significant is the contour function of the radar. In the contour mode, the areas of heavy precipitation are electronically eliminated to produce a dark hole on the screen, surrounded by the luminescent areas of lighter precipitation. One of the techniques in using airborne weather radar is in recognizing (and avoiding) areas where the line of light rainfall between the no-rain and the blacked-out areas (heavy rain) is fine. That indicates a steep rainfall gradient, characteristic of a thunderstorm. According to the manufacturer of this radar, areas where the rainfall rate is from one half to one inch per hour (or more) would appear as heavy rain. The rainfall rate exceeded a half an inch per hour over a broad area in front of the airplane. Then the airplane flew through an area of very steep gradient where the estimated rainfall rate increased rapidly from light rain to a rate of as high as, or higher than, 5 inches per hour. This extreme rain would have been within the general area of heavy rain and wouldn't have been specifically shown.

The NTSB surmised, "Given the high intensity precipitation levels of the storm and the comparatively short distance between the aircraft and the higher intensity precipitation levels of the storm, the aircraft's radar clearly should have shown a contour hole. However, since the aircraft was in rain at the time, the aircraft's radar might have been affected by attenuation to the extent that, when combined with the steep gradients associated with . . . the storm, the contour hole was distorted and interpreted by the captain as an area free of precipitation. The captain's comment, 'All clear left approximately right now,' seems to confirm this possibility, because the aircraft's course was then altered to the left, through the steep gradient and into the highest intensity level of the storm." See Figure 11 for the aircraft's path through the storm.

The crew neither asked for nor received any information from the controllers about the weather in the Rome area. They had no definitive information on severe weather that was there in spades, and they tangled

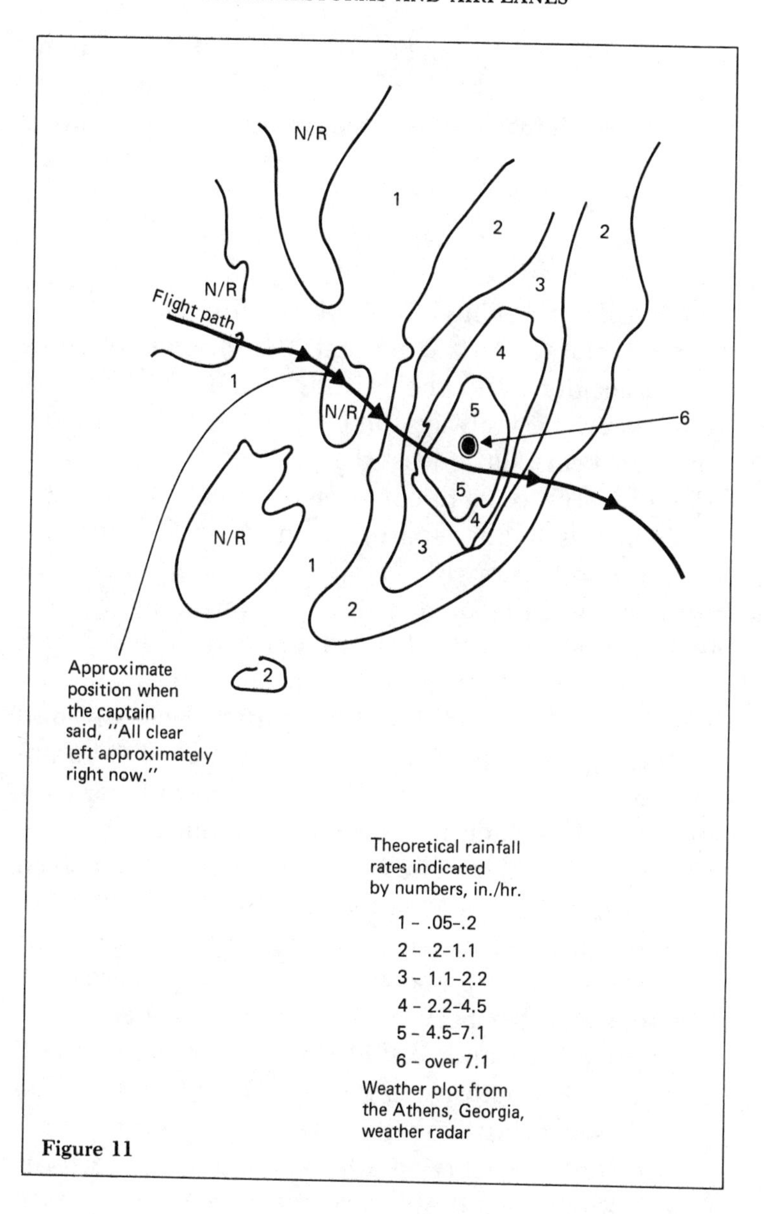
N/R
1
2
2
3
N/R
Flight path
4
1
N/R
5
6
5
4
N/R
3
1
2
2
Approximate position when the captain said, "All clear left approximately right now."
Theoretical rainfall rates indicated by numbers, in./hr.
1 - .05-.2
2 - .2-1.1
3 - 1.1-2.2
4 - 2.2-4.5
5 - 4.5-7.1
6 - over 7.1
Weather plot from the Athens, Georgia, weather radar

Figure 11

with conditions about as bad as they get. It was a unique and a tragic accident reflective of both the strength of nature and the weakness of the weather dissemination system.

SPEARED AND BURSTED

The crash of a Boeing 727 at John F. Kennedy International Airport in New York, and the subsequent study of the accident and the existing weather conditions, show that there is always something new to learn about thunderstorms. Or at least that there are always interesting new theories to consider.

When this flight departed from New Orleans and headed nonstop for JFK, the National Weather Service terminal forecast called for thunderstorms and moderate rainshowers in the JFK area after 6:00 P.M., after the proposed arrival time. The airline forecast for the New York area predicted just scattered clouds until 8:00 P.M., after which thunderstorms and light rainshowers would be possible. The NWS forecast was later amended to include heavy thunderstorms in the area. There was nothing to indicate that the crew received this revised forecast, even though it was issued well before they arrived in the New York terminal area. At 3:26 P.M., as the airplane was nearing the JFK area, the National Weather Service office in New York issued a strong wind warning calling for surface winds to 50 knots in thunderstorms in the New York City terminal area. This warning was distributed to the tower and approach control facility at JFK as well as to Eastern Air Lines operations at the airport, but there is no evidence that the warning was disseminated to pilots in the area.

The recorded automatic terminal information service at the airport was advertising good conditions as the flight approached: "Sky partially obscured, estimated ceiling 4,000 broken, 5 miles with haze . . . wind 210 degrees at 10 . . . expect vectors to an ILS runway 22 left . . ." There was no mention of thunderstorms. At 3:52:43 the controller transmitted, "All aircraft this frequency, we just went IFR with two miles, very light rainshowers and haze." Again, there was no mention of thunderstorms.

About five minutes later, as they were being vectored for the approach, the crewmembers of the 727 discussed the problems associated with carrying minimum fuel loads when confronted with delays in terminal areas. One remarked, "One more hour and we'd come down whether we wanted to or not."

At the same time, the National Weather Service radar at Atlantic City showed an area of thunderstorm activity along the northern edge of JFK airport. The area was 30 to 35 miles long and about 15 miles wide. Several groups of cells in the area had tops in excess of 50,000 feet. The largest group of cells was moving to the east-southeast. Also at about the time of crew discussion about delays, a Lockheed L-1011 abandoned its approach to runway 22L and reported, "We had . . . a pretty good shear pulling us to the right and . . . down and visibility was nil, nil out over the marker . . . correction . . . at 200 feet it was nothing."

A DC-8 pilot, who completed his approach just before the L-1011 went around, also had some problems and verbalized them to the controller after landing: "I just highly recommend that you change the runways and . . . land northwest. You have such a tremendous windshear down near the ground on final." The controller responded that they were indicating wind at 15 knots right down the runway when the flight landed, and the DC-8 pilot repeated his statement about the strong windshear and his suggestion that they change the landing direction to northwest. When the controller gave missed-approach instructions to the L-1011, he asked if wind was a problem and that flight answered, "Affirmative."

Two aircraft, another DC-8 and a Baron, followed the L-1011 and preceded the 727. The pilots of these aircraft stated that they suffered significant airspeed losses and increased rates of descent, but both had increased airspeed substantially and were able to cope with the problem. Neither pilot reported the windshear conditions; one stated that he didn't report it because it had already been reported and he believed that the controllers were aware of the situation.

The L-1011 crew's windshear report to the controller was recorded on the 727's cockpit voice recorder. While it was being transmitted, the captain of the 727 said, "You know, this is asinine." An unidentified crewmember added, "I wonder if they are covering for themselves." The controller then asked the 727 crew if they had heard the report from the L-1011 and they replied, "Affirmative."

The flight was cleared for an ILS approach to 22L when five miles from the outer marker, and at 4:00:54.5 the clearance was acknowledged. Almost a minute later the first officer, who was flying the aircraft, called for a completion of the final checklist. While this was being done, the captain stated that the radar was "up and off . . . standby." Then he added, "I have the radar on standby in case I need it, I can get it off later."

At 4:02:42 the controller asked the L-1011 crew if it would classify the windshear as severe. The response was "Affirmative."

A few seconds later the first officer of the 727 said, "Gonna keep a pretty healthy margin on this one." Another crewmember answered, "I . . . would suggest that you do," and the first officer responded, "In case he's right."

Control of the flight was handed off to the tower, and at 4:03:44 JFK tower cleared the 727 to land. The crew asked about the braking action, and the tower replied that the first half was wet but that they had no adverse reports. The following conversation ensued as the 727 moved downward and onward toward the runway:

UNIDENTIFIED: You got 500 feet.

(Sound of windshield wipers going to high speed.)

CAPTAIN: Stay on the gauges.

FIRST OFFICER: Oh, yes, I'm right with it.

FLIGHT ENGINEER: Three greens, 30 degrees, final checklist.

CAPTAIN: I have approach lights.

FIRST OFFICER: Okay.

CAPTAIN: Stay on the gauges.

FIRST OFFICER: I'm with it.

CAPTAIN: Runway in sight.

FIRST OFFICER: I got it.

CAPTAIN: Got it?

FIRST OFFICER: Takeoff thrust.

One point two seconds later the aircraft struck an approach light tower 1,200 feet short of the middle marker.

Five witnesses described the weather conditions when the 727 passed overhead as heavy rainfall with lightning and thunder, wind blowing hard from directions ranging from north through east. Persons driving on the boulevard on which the airplane finally came to rest said a driving rainstorm was in progress. One person in the north area of the airport stated that a violent wind was blowing from the northwest.

The approach was more or less normal up to the point where the captain said for the second time, "Stay on the gauges." The first officer responded that he was with it. But it was at about this time, as the airplane was descending through 400 feet, that the rate of descent increased from an average of about 675 feet per minute to 1,500 fpm and the aircraft began to move rapidly below the glideslope. Four seconds later the airspeed decreased from 138 to 123 knots in 2.5 seconds.

The 727 continued to deviate below the glideslope, and when the captain said, "Runway in sight," the airplane was only 150 feet high.

In a detailed analysis, the National Transportation Safety Board established a wind model for the 727's flight path. Using flight recorder information, it was established that the 727 probably encountered an increasing headwind to begin with. The wind changed from about a 10-knot headwind at 600 feet to a 25-knot headwind at 500 feet. An increasing headwind will cause a momentary increase in airspeed. All to the good to that point.

At about the time the 727 descended through 500 feet, it encountered a downdraft with peak speeds of about 16 feet per second (960 feet per minute), according to the NTSB. The headwind diminished to about 20 knots as the aircraft descended to 400 feet, where the speed of the downdraft abruptly increased to about 21 feet per second (1,260 feet per minute) and the headwind suddenly decreased from 20 to 5 knots over a four-second period. (A rapidly decreasing headwind will cause a momentary decay in airspeed.) It was here that the aircraft deviated rapidly below the glideslope. There wasn't much else that could happen, with the aircraft in the approach configuration and with the rate of descent increasing rapidly and the airspeed decreasing rapidly.

The wind model was considered consistent with the downdraft and outflow activity of strong thunderstorms. Close examination disclosed transient periods in which the combination of downdraft and decreasing headwind could have exceeded the aircraft's performance capability. That is, during these periods the aircraft could have continued losing altitude or airspeed, or both, even with maximum thrust and regardless of flight control inputs.

Nothing happened here that couldn't be associated with an "average" thunderstorm. And the situation could have been avoided by using standard thunderstorm avoidance techniques. The use of airborne weather radar to avoid cells, including those on final, would have worked fine, but the 727's radar had been turned off. There were, however, some unusual characteristics to the thunderstorms that moved across JFK airport on the afternoon of the accident. These characteristics were studied by Dr. T. Theodore Fujita of the University of Chicago, and a condensation of his research paper was published by Eastern Air Lines. In his paper Dr. Fujita brought some dramatically descriptive words to the thunderstorm subject.

The growth rate of the thunderstorm was in its peak when the accident occurred. The radar echo of the storm appeared as a "spearhead," moving faster than any other echo in the vicinity, and hidden in the spearhead echo were four or five cells of intense downdrafts, or "downburst cells." The activity developed on a weak front in the area.

The spearhead echo was defined in the paper as a radar echo with a pointed appendage extending toward the direction of the echo motion. The appendage moves faster than the original, parent echo, and during the mature stage the appendage turns into the major echo while the original loses its identity.

The downburst cells found in the JFK thunderstorm were different from most downdrafts. They moved rapidly and maintained a very strong downward current near the surface. There can be more than one downburst located in a spearhead echo, which will extend from the updraft cell in the direction of fast-moving, low-humidity air at the top of the storm. This high-altitude air is drawn into the top of the storm to support the downburst. It appears that a spearhead echo might have a relatively short life. The spearhead on the JFK storm reached its mature stage in about 50 minutes, and unfortunately for the passengers and crew of the 727, at about the time the aircraft was on the approach.

It is only natural for pilots to look for something unusual about a storm that bests a professional crew and a good airplane. Perhaps some comfort is found in learning of the spearhead echo and downburst and being assured that this characteristic is seen in only a small percentage of thunderstorms. But a 727 of the same airline survived a similar encounter at Atlanta Hartsfield airport after finding plenty of trouble in a storm that wasn't of the spearhead shape. The Atlanta incident is pertinent to the accident just described and survival at Atlanta might have been based more on the flying procedures used than on any difference in the storms.

The crew had thunderstorm forecasts when they left Indianapolis bound for Atlanta. At 2:54 on that August afternoon, the Atlanta weather reflected the first mention of a thunderstorm at the airport, and the automatic terminal information service was updated to include the storm. Information on a gusty surface wind was inadvertently omitted from the broadcast.

About 18 minutes before the 727 was on short final, the temperature at the airport dropped by 10 degrees. At 3:08 the National Weather Service radar at Athens, Georgia, showed numerous thunderstorms in the Atlanta area, including a comma-shaped echo over the approach course to runway 27L, the one the 727 would be approaching.

A DC-9 captain, also on approach, described three

cells north of the airport as well as a big cell south of the approach course to runway 27L. His airborne weather radar showed heavy rain in all these cells.

The crewmembers of the 727 said they observed scattered thunderstorms in the vicinity of the airport, both visually and on radar. The captain said that he selected the contour mode on his radar to examine the cells while inbound to the outer marker for runway 27L. He said he was not concerned with three little cells to the north; he was more concerned about the cell to the south.

At 3:10 the flight reported over a fix 8.5 nautical miles east of the runway threshold and was cleared to land. The controller added, "The winds are calm, and keep your speed up as long as feasible on final, sir. You'll break out of that rainshower in about three miles, and there is rain down the middle of runway 27 left right now."

This was acknowledged, and the captain said that he monitored the communications between the controller and two aircraft ahead of his on the ILS.

The 727 intercepted the glideslope outside the marker and was placed in the landing configuration at the marker. The reference speed for the approach was 120 knots; however, the first officer, who was flying the aircraft, said that he attempted to hold 135 knots.

The crew said the ground was visible when the 727 passed over the outer marker and that there was light rain, light turbulence, and a little airspeed fluctuation. When the airplane was 1,000 feet above the ground, the turbulence became moderate and the rain "heavy."

Simultaneously with the increase in rain and turbulence the indicated airspeed began to fluctuate. It decreased from 135 to 120, then increased to 140 and, a few seconds later, decreased to between 108 and 110 knots. When the airspeed decreased, the first officer saw that the rate of descent had increased from a rate of from 500 to 700 feet per minute to 1,000 feet per minute. When 800 feet above the ground, the first officer rotated the aircraft to a 10-degree nose-up attitude and called for takeoff power.

According to the first officer, the pitch correction and added thrust had no effect. The rate of descent increased to 1,500 feet per minute and then to 2,000 feet per minute. The first officer then rotated the 727 to a 15-degree nose-up attitude and pushed the thrust levers to their forward stops. The captain insured that the levers were all the way forward.

When the aircraft was between 500 and 700 feet above the ground, at an airspeed of between 105 and 110 knots, the stall warning stick-shaker and the ground proximity warning systems activated, the below-glidepath light illuminated, and audio "pull-up" and whooper warnings began.

When the stick-shaker activated, the first officer said that he reduced the pitch attitude to 12 degrees and it stopped. The captain felt that the first officer had "overreacted" and told the first officer to pull the nose back up. The flight engineer said that the rate of descent was 2,100 or 2,200 feet per minute at this point. The procedure of letting the airspeed decay to just above stick-shaker speed is accepted in such encounters and is practiced in the simulator.

According to the crew, the aircraft flew out of the precipitation when 375 feet above the ground. At that time they were able to arrest the descent and transition to a climb.

The local controller told the flight they had received a low-altitude alert and asked if they had the airport in sight. The captain replied, "No, sir, we kinda missed out here. There's quite a bit of rain . . . a windshear out there. I don't see how anybody could make an approach to the left one."

When the power levers were advanced fully, the engines were developing more than their rated power, and rpm and exhaust gas temperatures exceeded the redlines. The engines operated satisfactorily during the go-around and the 50 minutes the 727 flew while holding for better weather and another approach into Atlanta.

There was a stationary front north of Atlanta this day, with nothing on the map to indicate severe thunderstorm activity. It was more an unstable summer afternoon than anything else. The air carriers chose to operate closer than five miles to cells containing heavy rain; it is reasonable to expect some encounter with the dynamics of thunderstorms when operating in that manner.

It is interesting to compare these two encounters. The first attracted a great amount of attention because the airplane crashed. The thunderstorms in the system were undoubtedly severe, and the spearhead echo and downburst relationship to the accident is interesting. But in the second case the storm and resulting downdraft appear to have been stronger than the one in which the 727 was lost. At the Atlanta airport a steady wind of four knots with a peak gust to 32 knots was recorded in a thunderstorm. The peak gust of a storm is said to be a product of the storm's movement over the ground and the downdraft. These storms were not moving fast (the steering currents at the 18,000-foot level were very weak), so most of the gust had to be coming from the downdraft. If the gust was 28 knots, that's 48 feet per second, or 2,880 feet per minute. And some of its strength was probably lost to surface friction. Certainly the 727 crew's inability to keep the rate of sink below 2,000 feet per minute even when pulling in excess of rated power is a tribute to the strength of the storm.

By comparison, the estimate of downburst strength in the accident-related storm was 21 feet per second or 1,260 feet per minute. This was closer to the ground, but the Atlanta downdraft might well have been that strong or stronger as it fanned out and changed from a downdraft to a wind shift at lower altitudes. In other words, the Atlanta storm might have been significantly stronger than the JFK storm. It didn't make the headlines because the crew managed to avoid an accident.

Another point to consider is turbulence. The Atlanta airplane was bumped around a little. When pilots fly into turbulence it probably causes an increase in the instrument scan and a heightened awareness of what's going on. The smoother onset of a downdraft thus might not be caught as quickly as a rough onset. Too, the runway was visible to the JFK crew and at least one crewmember might have been concentrating on that, leaving the instrument flying to the person operating the controls. In the Atlanta incident, it is apparent that everyone (including the flight engineer) was interested in the tale told on the instrument panel.

Since these events, a government-funded study offered the thought that heavy rain alone is enough to cause sink-rate problems for a heavy airplane in the approach configuration. In fact, the researchers said that the roughening effect of heavy rain on the airfoil can reduce lift by 30 percent at the high angles of attack characteristic of the approach or missed approach profile. Most investigations have centered on windshear and downdraft effects without regard to rain-induced problems; in the case of the 727 that crashed at Kennedy airport, these researchers (from the University of Dayton Research Institute) concluded that "Eastern 66 wind profiles may have been overestimated by a factor or two." Also, research at the National Severe Storms Laboratory suggests that downdraft strength close to the ground is not as strong as has been advertised in some accident reports.

Certainly the mechanics of a thunderstorm connect the downdraft with heavy rain, so the two do go hand in hand. The variation in wind must also be considered. It is highly logical (as well as technically correct) that rain would have some effect on aircraft performance. In heavy rain not associated with a thunderstorm this is evident even in a light airplane. Climb suffers or, at cruise, speed drops off a bit. The message, though, is that being low and slow in the heavy rain of a thunderstorm can lead to a triple dose of problems as the falling water, the downdraft, and the shifting winds go to work on the airplane.

Finally, the DC-9 lost near Atlanta, the 727 lost at JFK, the 727 lost at Denver, and the 727 nearly lost at Atlanta have one thing in common: all were being flown by the first officer. This has to be accepted as coincidence, but it is still an interesting point to ponder. The captain is in command, but does the captain allow the first officer to continue into conditions that he would avoid were he actually flying the airplane?

Turboprop Frontal Assault

The most common airframe failure in an area of thunderstorms comes after a pilot loses control of the airplane. While this happens almost exclusively to general aviation pilots, airline crews have not always been immune. An Electra that was lost in Texas in May 1968 is a graphic example of loss of control, with the cockpit voice recorder giving insight into the thought processes that might precede a thunderstorm encounter, and with the flight recorder creating a record of the dynamics of thunderstorm penetration and a loss of control. It's all as applicable to a single-engine retractable as to an airliner.

The Electra was en route from Houston to Dallas one afternoon, flying at flight level 200 when, at 4:35:53, the first officer said to the controller, "A few miles up the road we'd like to deviate to the west, looks like there is something in front of us." This was followed by a handoff to Fort Worth center. The request for a westerly deviation was repeated, along with a request for a descent to 15,000 feet. The Fort Worth controller replied, "Suggest deviation east of course, the aircraft are deviating that way at the present time." The flight replied, "Just a little bit to the west would do us real fine."

An even lower altitude was requested but was not immediately available because of other traffic. At 4:37:23 the captain said, "It looks like there's a hole up ahead to me." The first officer agreed, and shortly thereafter the captain made an announcement to the passengers that they would be arriving in Dallas on time, that there was a "little line of thundershowers" ahead, and that they would be deviating west for a better ride.

At 4:41:07 the captain advised the passengers that he was turning on the Seat Belt and No Smoking signs "in the event it's a little choppy in the area." He announced that the radar was working and that they would be able to "go well under and to the west of all the thundershowers."

Next the captain said to the crew, "I guess I can go under." At 4:46:30 the captain told the first officer to ask the controller for any reports of hail in the area, and the controller replied, "No, you're the closest one that's ever come to it yet. . . . I haven't been able to, anybody to, well I haven't tried really to get anybody to go through it, they've all deviated around to the east." Following this transmission the captain said to the first officer, "No, don't talk to him too much. I'm hearing his conversation on this. He's trying to get us to admit . . . big mistake coming through here."

The first officer next said, "It looks worse to me over there." The gear warning horn sounded (which it would do when engine power was reduced below a certain level) and the captain said, "Let it ring."

At 4:47:20 the captain said, "Let's make a one eighty." The first officer called the controller and requested permission to do this; it was approved, in either direction. The captain rolled into a right turn. At 4:47:29 a sound similar to hail or heavy rain was recorded. The captain said, "Let me know when we come back around there to reverse heading for rollout," and at 4:47:32.5 the first officer said, "Three forty." (He was perhaps referring to airspeed, because the captain was turning to a southerly heading.) Immediately afterward, the landing gear warning sounded again and the captain said, "Right." At 4:47:41.93 a fire warning bell was heard, and about half a second later a sound appeared on the tape that was described as being similar to the noise of an in-flight breakup.

Witnesses in and around the area reported that it was raining or hailing, or both, with high winds and lightning at the time of the accident. The clouds were described as being dark, black, green, or purple, and some witnesses reported a rolling or boiling motion in the leading edge of the clouds.

This active squall line was moving ahead of a cold front, but weather radar reports showed that it was not solid, and indeed, flights had been deviating to the east, as suggested by the controller, with no trouble. Statements from flightcrews operating in the area emphasized the severity of the storm area that the Electra crew attempted to penetrate, with some mention of a green appearance to the storm and with tops estimated to be from 26,000 to 40,000 feet. The conditions were adequately covered by forecasts and sigmets.

The flight recorder readout showed a smooth and normal flight until 4:47, at which time excursions began in the g-tracing along with some sharp, small changes in the heading indication. The g-excursions continued to increase until reaching a maximum of 4.3 at about the time of breakup. A major spike in the trace, to 3.8 positive g, occurred about four seconds before the 4.3.

Concurrent with the increase in g-loading, the airspeed increased abruptly from about 200 knots to about 360 knots over a period of eight seconds. While this was happening, the heading changed from 350 degrees to 184 degrees in under 20 seconds.

There were some errors in flight recorder information caused by extreme angles of bank (in excess of vertical) and other factors, but it still offered a clear picture of a lateral upset followed by a longitudinal upset.

This sequence of events—a rapidly steepening turn with increasing airspeed and g-loading—is probably what precedes most general aviation thunderstorm accidents. Once the decision is made to turn around or turn away from the storm, the temptation is to do it quickly. Too, the noise of rain or hail pelting the airframe can be distracting and can actually draw the eyes to the point of the most perceived noise—the windshield. Turbulence could only enhance the possibility of a loss of lateral control in such a situation. In this case, the NTSB determined that the airplane probably broke up under the following conditions: an airspeed of 330 knots, a heading of 200 degrees, an altitude of 6,750 feet, and a *descent rate of 480 feet per second.* That descent rate is equal to 28,800 feet per minute, which gives some idea of the attitude at the time of the breakup, and strongly suggests that recovery would have been impossible before hitting the ground even if the airplane had not broken up. That is probably true in most of the general aviation thunderstorm accidents. An airframe failure before ground contact is not a cause of the accident but one of two possible endings in an accident that became inevitable when control of the aircraft was lost.

Why might this captain have stuck to his guns on the westerly deviation for so long?

The radar antenna cockpit control recovered in the wreckage was in the 8-degree tilt-up position. If the antenna had been at this setting as the aircraft approached the squall line, lower cloud tops between cells could have been interpreted as light spots between cells. This could have continued misleading the crew even as the aircraft approached the storm. Also, hail has less radar reflectivity than rain, and hail in the area could have combined with an incorrect tilt setting to make the picture on airborne weather radar look better than it actually was.

In the end, it was probably the visual appearance of the clouds and the precipitation that prompted the decision to retreat. In its report, though, the NTSB pointed out that the decision to reverse course was not in keeping with recommended company procedures. "Once in an area of turbulence," the report stated, "the crew is expected to maintain the attitude of the aircraft as nearly straight and level as possible, and maneuvering is to be kept to a minimum until the turbulent area is cleared. The possibility of gusts being added to control inputs, and resulting in an upset, is a consideration that must be assumed by the pilot." In this case, the turbulence-induced g-traces were far from strong enough to damage the airframe, and the airplane might have made it through the storm had its crew continued straight ahead.

The Pennsylvania Storm

Pilots who live in the northeast and fly westward a lot, and pilots who do the reverse, will testify to the weather over the Appalachian Mountains. "If you are going to get beat up, it's probably going to be between Pittsburgh and Harrisburg." Fronts tend to move slower over the rough ground, and even when there is no identifiable front, buildups can be numerous over the mountains. A high-pressure area north of the New York area can promote an easterly flow that is lifted a bit as it reaches higher terrain. In many situations, it collides with a southwesterly flow coming from west of the mountains. Given some instability aloft, this is enough to get things going.

There was a rather complex and slow-moving array of frontal activity in the area as an Aztec proceeded from Indianapolis toward central New Jersey early one June day. A low-pressure system was located in western Maryland with one cold front extending southwestward from southeastern Pennsylvania and another oriented east–west across Pennsylvania, just north of Harrisburg. The air was unstable, and the synopsis suggested the possibility of waves developing on the front as it drifted southward. Thunderstorms were reported over the Appalachians before the pilot reached the area. A Pittsburgh radar weather observation reported 60 percent coverage of very strong echoes, increasing in intensity, from 120 miles northeast of Pittsburgh around to 130 miles southwest of Pittsburgh. The highest tops were to the southeast and east, with a strong echo, top 42,000 feet, 144 miles to the east.

That it was a day for convective activity was apparent early in the flight of this Aztec. There was conversation with a controller about a buildup when the airplane was in the vicinity of Dayton, flying at 7,000 feet. There was further conversation about weather when the airplane was passing Wheeling, West Virginia, still to the west of the mountains. The pilot reported that he was flying in precipitation and said that it was getting a little bumpy. Then he reported being out of the rain, between layers, and in smooth air. When the Aztec was 10 miles west of Wheeling, the controller told the pilot, "There's some more precipitation you may encounter, sir. I have no idea what's in it. It's, ah, relatively scattered, ah, on my radar. It may be just about the same as what you went through, but you'll be talking to Pittsburgh approach shortly and they'll be able to give you more information on how the ride is."

As the aircraft moved eastward, there was nothing pertinent regarding weather as it flew by Pittsburgh. The Aztec did climb from 7,000 to 9,000 feet, and by the time it was passing Altoona, Pennsylvania, the flight conditions had clearly worsened.

PILOT: Ah . . . what type of weather do I have ahead of me? It's pretty rough in here.

CONTROLLER: Well, sir, I'm painting weather all along Victor-12 clear from ah, oh, about 20 miles east of Allegheny [a VOR near Pittsburgh] eastbound, ah, to about your twelve o'clock position about 10 miles. After that it should be in the clear.

Control of the flight was then handed off to New York center.

The first order of business was a new clearance for the Aztec, to conform to standard arrival routing into the New Jersey area. The controller also broadcast notice of a sigmet but did not say what area or phenomenon was involved.

Clearance business out of the way, the pilot asked the controller the following question at 5:02:20 P.M.: "Do you have a cell right in front of me?"

CONTROLLER: . . . say again.

PILOT: What's the weather directly ahead of me?

CONTROLLER: I have some weather painted out in front of you. I really don't know exactly what it is, er, some airplanes have been going around it, some have gone through it, er, I really can't tell. All I have is a couple of H's on my map that, er, really don't tell me too much.

PILOT: . . . vector me around the other way. I'm in a hail storm.

CONTROLLER: . . . I suggest, er, a right turn, a right turn, heading 180, er, I have no idea what the weather is to the south of that, but it looks like the quickest way out.

PILOT: Nine nine one is in a severe storm. Get me around the other way.

CONTROLLER: Maintain a heading of 180. That should bring you out pretty quick.

PILOT: One eighty. . . . Center, am I inverted, over?

CONTROLLER: I'm sorry, 991, say again.

PILOT: I'm diving.

The pilot then reported being in a spin as the controller continued telling him that the best he could do was advise a southerly heading. Then came momentary good news: "Nine ninety-one is in level flight, I see the sun to my right."

The controller then cleared the Aztec pilot to deviate any way and asked for an ident from the transponder. Then trouble started anew.

PILOT: Nine nine-one is, I'm showing zero on this thing. . . . Nine nine one is in a spin.

CONTROLLER: I show you south of the storm now, south of the storm, er, are you in level flight, over?

The pilot answered the controller, but the message was garbled and was heard only by another aircraft in the area.

A ground witness, a pilot, located about 20 miles west of Harrisburg reported hearing an airplane high in a storm that had just reached his location with high winds and rain. He detected that something was wrong and ran outside, where he heard what sounded like an airplane coming straight down out of the storm. He followed the sound into the mountain that was directly in front of him, about a mile away. There was an explosion when the airplane hit. He couldn't see the impact because of heavy rain, but 10 or 12 minutes later, when the rain abated somewhat, smoke was visible. This witness reported that the engines were at or close to full throttle as the airplane descended.

The Aztec's airframe failed in flight, with the stabilator apparently the first thing to fail. The farthest piece was 900 feet from the main wreckage, so the altitude at which the failure occurred was probably well below its assigned cruising altitude.

This was a particularly strong thunderstorm system. There were reports of baseball-size hail, and recorded wind gusts reached 65 knots. The official report at the airport documented hail that was three quarters of an inch in diameter. In its finding on the accident, the NTSB reported that the weather was "slightly worse than forecast."

The controller reported to the pilot that he was south of the weather, but the witness's statement suggests that he wasn't quite out of the precipitation when he crashed. It was close, but a miss is as good as a mile.

When moving from the back of a weather system to the front, we often save the worst for last. Storms are often meaner on the side toward which they are moving, and on the side from which they get their major feed of moisture. So the pilot of the Aztec was probably facing a steadily worsening situation as he flew eastward. He apparently lost and then regained control of the airplane, only to lose it again. Based on a study of the NTSB report, it does not appear that the airplane suffered any failure as a result of a turbulence encounter at cruise; rather, it failed after the pilot lost control for the second time and the airspeed probably exceeded the design limit. Based on the dynamics of thunderstorm development, the airplane was in or near an area of maximum turbulence when all this took place.

The H symbols the controller referred to on his radar screen are those used to depict heavy precipitation. The controller indicated to the pilot that these didn't tell him too much, but in this case they did prove to be indicative of severe weather activity.

SQUALL LINE ENCOUNTER

Small-airplane airframes seldom fail on the initial encounter with thunderstorm turbulence. We've seen that a loss of control usually precedes the failure. Failures of larger airplanes because of turbulence alone are even rarer, because an airplane's response to a gust, and the g-load encountered, is at least partly proportional to wing loading and the larger ones have higher wing loading than light airplanes. But the airframe of an airline BAC 1-11 twin-engine jet did fail as it approached a squall line in August 1966. The accident occurred in Nebraska, and the circumstances and research done in relation to this event are revealing.

The accident happened at night, with a nearly full moon visible to the southeast of the squall line. The BAC 1-11 departed from Kansas City at 10:55 P.M. en route to Omaha and was cleared to climb to flight level 200. There was discussion with the controller about weather, though, and the crew asked for and received clearance to fly at a lower altitude. Initially they leveled off at 6,000 and were later cleared down to 5,000.

The crew talked with the controller about the weather ahead as it was displayed on his radar and exchanged weather information with another flight that had departed from Omaha and was headed southeast. The crew of the other flight advised of light to moderate chop from 15 southeast of Omaha and said that the picture on their radar suggested they would be through the weather in another 10 miles. This conversation was terminated at 11:08:30 and was the last transmission from the BAC 1-11.

As the crew flew toward the line, it had good information on the intensity of the system. Even before takeoff in Kansas City, the captain of another flight told the captain of this one that there was a "solid line of very intense thunderstorms with continuous lightning and no apparent breaks, as long and mean a one as I had seen in a long time, and I didn't feel the radar reports gave a true picture of the intensity."

At 11:04:44 the crew had requested permission to deviate left of course. The center controller told them that the line appeared "pretty solid all the way from west of Pawnee to Des Moines." Along with the discussion with the other flight, there was intermittent cockpit conversation regarding a deviation from 11:07:08 until 11:10:59. Then, "Ease power back" were the last words from an intelligible crew voice on the cockpit voice recorder tape. Eleven seconds later, a noise started in the cockpit and increased rapidly to a constant level. The sound was described as a "rushing air" noise. This was followed by another unidentified sound, an electronic flutter sound, warning horns, and then by the termination of the tape at ground impact.

The airplane crashed at 11:12. Witnesses reported that they saw the aircraft approach and fly into or over a shelf of clouds preceding the line of thunderstorms. The clouds were described as rolling in a circular motion from top to bottom. Shortly after the accident, the wind shifted from light southerly to strong northerly with velocities as high as 60 miles per hour reported. Rain began shortly after the accident. All the witnesses were sure that the aircraft did not penetrate the main line of thunderstorms, and the flight was 5 miles away from the nearest echo observed on traffic control radar when its target disappeared from the scope.

The investigators surmised that the airplane was flying basically straight and level when it was subjected to violent forces that caused it to accelerate upward and to roll left. At this time, the right tailplane and the fin failed. The aircraft then pitched nose down, and in a second or two the right wing reached its negative ultimate load. When the wing failed, the rupture of a fuel tank released fuel, which ignited, creating the ball of fire observed by witnesses.

At the time the encounter began, the aircraft was at the proper airspeed and configuration for flight in turbulence. However, the horizontal component in the gust that was encountered resulted in an increase in airspeed and probably accounted for the "rushing air" noise on the cockpit voice recorder. The ambient noise level on the cockpit recorder before the encounter was reproduced in a test flight by flying the airplane at an indicated airspeed of approximately 270 knots, the correct value for turbulent air penetration. It was further determined that the "rushing air" noise on the tape could be reproduced by increasing the airspeed 45 or 50 knots, especially if the aircraft had a large sideslip angle.

How strong a gust would be required to break an airplane like a BCA 1-11?

The regulations under which this airplane was certificated required that the airframe withstand gust velocities of 66 feet per second. The design parameter is based on an "artificial gust" of a specific shape which, when used in the appropriate formula, will produce accelerations in line with those which have been measured on similar aircraft in similar weather conditions. (By comparison, the design gust limit for older general aviation aircraft is 30 feet per second; it's 50 fps on newer airplanes—see Chapter 4.)

Studies showed that the gust encountered by the BAC 1-11 was much stronger than the one covered by certification requirements. It was calculated, for example, that a 140-fps gust applied from the right and angled upward 45 degrees would be required to fail the fin and tailplane as it happened in this accident. (As a point of reference, 140 fps is almost 85 knots.) The effect of that 140-fps gust was calculated at 300 knots indicated airspeed; it would have taken a 158-fps gust to cause the failure at 270 knots. Put another way in the report's calculations, the maximum gust the aircraft could withstand would range from 66 fps at 270 knots to 50 fps at 320 knots.

A number of other flights operated through or in the

vicinity of this squall line at about the same time and their flight recorders were examined. (The flight recorders in use at this time recorded g-load, airspeed, heading, and altitude. The recorder on the accident aircraft was lost in the ensuing fire. Only its cockpit voice recorder provided useful data.)

A Convair 580 twin turboprop flew through the area at 3,000 feet using radar and visual guidance furnished by lightning. The flight originated in Omaha and was southeastbound. The captain reported clearly defined cloud bases at 3,500 to 4,000 feet and almost continuous lightning. Then the cloud base lowered, the flight went on instruments, the turbulence increased to moderate "plus," and heavy rain was encountered. The Convair then moved into clear air southeast of the storm, where it encountered severe turbulence. The g-loads recorded were a maximum of +1.9 g to 0 g (the aircraft is at 1 g in unaccelerated flight) while moving through the squall line. One single large excursion of plus 2.85 was recorded after the flight was in clear air southeast of the line. The airspeed trace varied from 8 to 10 knots.

The other aircraft was the one with which the crew conversed, also a BCA 1-11. That aircraft, flying from Omaha to Kansas City, deviated 40 miles to the east for penetration of the line. The crew used a combination of airborne weather radar and ground radar information and completed its penetration 29 miles east of the accident site about ten minutes after the crash. This aircraft recorded vertical accelerations ranging from +2.5 g to −0.3 g, with the most extensive excursions occurring when it was through the line, on the southeast side.

A third BAC 1-11 crew looked at the squall line and elected not to penetrate.

The 2.5 g encountered by the BAC 1-11 that penetrated and survived correspond with the limit load factor of the airplane. If this 2.5 g and the 66-fps gust design rule seem low, consider that the means of setting criteria for withstanding atmospheric turbulence evolved over a lot of years and is based on experience gained by monitoring operations. NASA's flight recorder program has generated plenty of information on flight loads encountered in normal operations. The data on turbine-powered commercial aircraft were based on millions of miles of experience at the time of the accident, and the probability of an airliner encountering a 66-fps gust was calculated at once for every 2.78 million nautical miles flown. Further, data indicate that the design gust will be encountered once every 1,820 miles during actual flight within thunderstorms. That's a lot of miles worth of thunderstorms—especially if the crew follows guidelines of thunderstorm avoidance.

The conclusion was that the turbulence encountered by this flight was of such magnitude that it would have caused the airframe of any airline aircraft to fail. Given this, and given the fact that the crew was employing proper turbulence penetration techniques, there must be a lesson in the severity of the weather that was encountered.

A meteorologist testified at the NTSB hearing that the outflow wind speed of a storm system is proportional to the amount of surface rainfall if the height of the clouds is equal. In other words, harder rain equates with stronger wind and vice versa. In the case of this squall line, the rainfall variation from one location to another was unusually large. Rainfall in an area about 30 miles northeast of the accident site was over four times that to the west of the site. The rainfall variation can have a dramatic effect on wind-shift line and the velocity of the first gust. And strong horizontal wind-shear and accompanying eddies can develop between the areas of different velocities. Figure 12 shows the situation as it existed at the time of the accident.

The meteorologist witness said that at the time of breakup the airplane was in an area most favorable for the development of roll circulations with horizontal vortexes parallel to the wind-shift line. He also stated

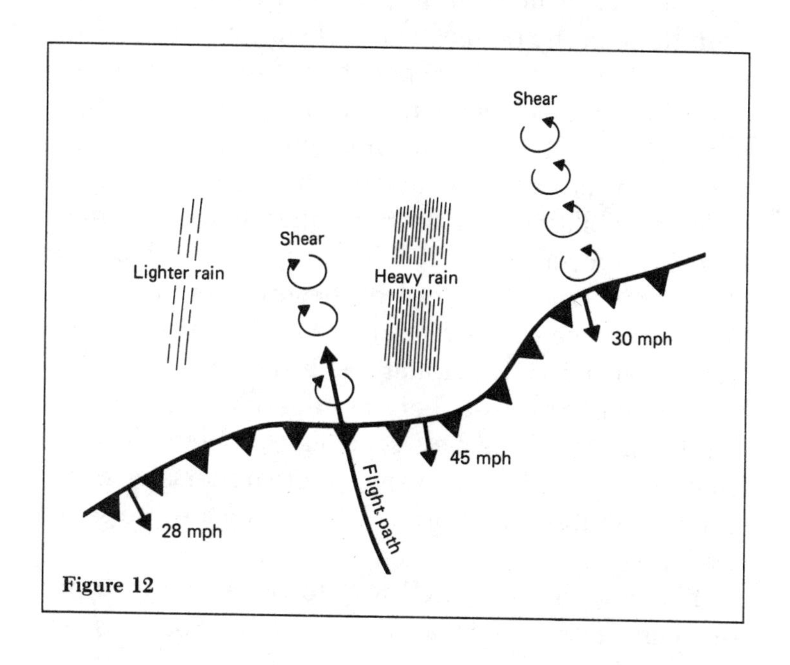

Figure 12

that horizontal windshear is strongest at levels between 2,000 and 3,000 feet above the ground. He concluded that the flight deviated to the left to avoid an area of intense radar echo, the heavy rain area, and flew into the area most favorable for the development of roll circulations.

The aircraft was initially operating in air that was flowing into the storm system. There was a tailwind component as it got near the storm, and the airplane flew into the air that was flowing out of the storm, or into a headwind. Going suddenly from a tailwind condition to a headwind condition produces an increase in airspeed. One has only to know the basics of circulation in and around a thunderstorm to visualize this.

It was a clear illustration that the meanest areas are often those in the path of a fast-moving storm. The two aircraft that successfully penetrated while flying toward the southeast encountered the worst turbulence after clearing the area of rain. This aircraft encountered destructive turbulence before reaching the rain.

The possibility of heavier rain equating with stronger wind is important, too. It tells us that in a squall line situation we might indeed be heading for the worst shear turbulence if the aircraft is flown just to avoid the areas of heaviest precipitation. Research at the National Severe Storms Laboratory doesn't exactly verify this. It has not established a relationship between rainfall rates and wind. However, researchers have recorded pulsations in storms, which indicate the same result.

There's really no "safe" way to penetrate a strong squall line. Some pilots say it's best to go through as low as possible, but this is where the strongest shear, horizontal activity, and roll circulation will occur as the downdraft fans out ahead of the storm. Others suggest going through at an altitude just above the roll cloud, as the BAC 1-11 crew apparently attempted to do, but the disturbance here is apparently a lot bigger than the roll cloud. In total it appears that there are times when the airplane should be allowed to wait patiently for better conditions.

The Harder They Fall

Prior to departure from San Juan, Puerto Rico, on December 8, 1963, the captain of the Boeing 707 was briefed on the weather along the route to Philadelphia. The briefing included discussion of a sigmet on thunderstorm activity and the forecast times of a frontal passage at certain east-coast cities. The flight then operated normally to Baltimore, a scheduled stop.

An airline operations representative at Baltimore talked with the captain about the weather on to Philly and provided him with copies of the 7:00 P.M. sequence reports. He told the captain that the front had passed Baltimore "a little while ago" and that it would be in Philadelphia at about 8:25.

The flight departed Baltimore at 8:24 and was cleared to 5,000 feet. It reported over the New Castle VOR at 8:42, where control was handed off to Philadelphia approach control. The controller there said, "Philadelphia weather, now 700 scattered, measured 800 broken, 1,000 overcast, 6 miles with rainshower, altimeter 29.45, the surface wind is 280 degrees at 20 with gusts to 30. I've got five aircraft have elected to hold until this . . . extreme winds have passed. . . . Do you wish to be cleared for an approach or would you like to hold until the squall line . . . passes Philadelphia, over?"

The crew elected to hold but a few minutes later advised Philadelphia that they were ready to start an approach. They were soon told to continue to hold, that they would be cleared as soon as possible. The crew acknowledged with "Roger, no hurry, just wanted you to know that . . . we'll accept a clearance."

Approximately eight minutes later, at 8:58:56, the following transmission was heard on the frequency: "Mayday, Mayday, Mayday. Clipper 214 out of control. Here we go." Seconds later the first officer of another flight in the same holding pattern 1,000 feet higher transmitted: "Clipper 214 is going down in flames." The first officer had seen the flight descending, on fire. He later described the weather in the holding pattern as cloudy, with comparatively smooth air. Ground lights were occasionally visible through breaks in the clouds. The crew of this flight observed a lightning strike on their aircraft while in the holding pattern, and later examination revealed lightning damage to the left wingtip and the empennage.

Seven ground witnesses stated that they saw lightning strike Clipper 214. Three others saw a ball of fire appear at the fork of one end of the lightning stroke. Many other witnesses observed lightning, an explosion, and objects falling from the aircraft.

Nearly 600 pieces of wreckage were outside the main crater in an area approximately 4 miles long and 1 mile wide. There were two distinct wreckage paths and three concentrations in this area.

One was a straight path and included the wreckage farthest from the main crater. It contained nearly all pieces of the left outer wing panel. The other was a slightly curved path consisting first of low-density items such as Seat Occupied cards, insulation, and fragments of thin skin and stringers. Further along were more dense pieces of wreckage. Most of the rest of the wreckage was in or near the crater except for the numbers one and four engines and their pylons and pods. These were from 1,440 feet to 1,925 feet from the crater and had separated from the airframe before ground impact. Separation was due to excessive load factors.

Multiple lightning-strike marks were found on the left wingtip, and other lightning damage was found on various pieces of the left wing.

The NTSB called a witness from the U. S. Weather Bureau (now National Weather Service) to testify at the hearing. Various technical reports were also reviewed. The accident report's coverage of this is interesting:

> These sources indicate that a lightning stroke begins when the air's resistance to the passage of electricity breaks down. At that time a faintly luminous stepped leader advances toward an area of opposite potential, the earth in the case of cloud to ground lightning. The difference in electrical potential between a cloud and the ground may be in the order of ten to one hundred million volts and discharge currents may exceed 100,000 amperes with 10,000 amperes per microsecond

(one millionth of a second) rate of current rise.

The stepped leader advances toward the ground in a series of discrete branching movements, forming an ionized path from the cloud. As the branch of the stepped leader is approaching the ground, the intensified electric field causes an upward moving streamer to form at a ground projection and advance toward the stepped leader. As the oppositely charged leaders meet, completing the ionized channel, an avalanche of electron flow follows, discharging the cloud to the ground. This entire sequence is accomplished in approximately one millisecond (one thousandth of a second). Additional charge cells in the cloud may then successively discharge through this main ionized channel as a single flickering flash which may last as long as one second. The electron flow suddenly heats the ionized channel to about 15,000 degrees C, expanding the air outward with a thunderclap. The discharge can also occur between oppositely charged regions within a cloud, or in different clouds.

If the stepped leader of a stroke approaches a flying aircraft, the intense electrical field induces streamers from the extremities of the aircraft out toward the approaching stroke. The stepped leader contacts one of these aircraft streamers, completing the ionized channel to the aircraft and raising the potential of the aircraft to the order of 100 megavolts. (One megavolt equals one million volts.) This high potential produces streamers from all the extremities and high gradient points of the

> aircraft. These streamers can have sufficient energy to ignite fuel vapors. Meanwhile, the stroke leader continues on from the aircraft to another cloud, or to the ground to complete the ionized channel for the electron avalanche.
>
> Statistics indicate that the majority of lightning strikes to aircraft occur at ambient temperatures near the freezing level. This correlates with thunderstorm electrification theories that charge separation occurs about the freezing level. N709PA [the aircraft lost] was at or near the freezing level just prior to the accident.

The National Transportation Safety Board found that the probable cause of this accident was lightning-induced ignition of the fuel-air mixture in the number one reserve fuel tank with resultant explosive disintegration of the left outer wing and loss of control. The tank referenced is a 434-gallon tank near the tip; it was virtually empty at the time of the accident.

Turbulence was ruled out. It was light to moderate in the area where the 707 was operating, and not of the strength normally associated with a loss of control or structural failure.

Lightning protection provisions have been improved since that 707 went down in a frontal area, but an even bigger airplane, a Boeing 747, was victim of a thunderstorm 13 years later, in 1976, and lightning again appears to have done the deed.

The 747 was owned by the Imperial Iranian Air Force and was near the end of a flight from Teheran to Madrid when it was lost.

The crew of the aircraft was aware of thunderstorm activity, and the captain remarked that the storm ahead would "tear us apart" if the aircraft entered it. It was daylight, so the clouds would have been easily visible as long as the aircraft was in VFR conditions. The flight was descending to 5,000 feet and receiving navigational information to steer around the activity. No sounds associated with turbulence were detected on the cockpit voice recorder tape before the beginning of what the NTSB classified as "significant" events.

The exclamation "We're in the soup!" was the first indication of a problem. Three seconds later an electrical transient occurred on the recorder tape. This was interpreted as an indication that the airplane had been struck by lightning. Next came a noise interpreted as thunder, then an explosion. Other abnormal sounds were recorded, and a member of the crew said, "Watch your autopilot." This was followed by the sound of the autopilot disconnect and a 24-second period of relative calm.

The captain then stated, "The flight control is not working." The gear warning horn sounded, and the recording ended in a loud noise 12 seconds later. The elapsed time from the electrical transient until the end of the recording was 54 seconds.

The aircraft wreckage was over an area about 5 miles long and 2 miles wide. Three deposits of wreckage were within the area. The first contained the left wingtip and substantial other parts of that wing. The second area was about 4 miles farther along and contained more parts associated with the left wing of the aircraft. The final area contained all four engines, the inboard section of the left wing, the fuselage, empennage, and the entire right wing.

This accident was not subjected to a full NTSB investigation, but that group did participate and issue a special investigation report. Instead of finding a probable cause, the report issued findings and plausible hypotheses.

It was determined that lightning struck the aircraft an instant before an explosion, and that damage to the wing in the area of the number one fuel tank was the result of a low-order explosion. Three fires and other wing damage were the results of explosions occurring in tanks in the left wing. Failures of structural components within the wing resulted in extreme engine oscillations and wing oscillations. The pattern of wreckage deposits was not compatible with gusting or turbulent wind conditions, and the observed damage to the wingtip could not have been caused by gust or aerodynamic loads. The report considered turbulence as a cause, but a lightning strike and the subsequent explosions in the wing appeared the more likely cause of the failure.

Those two accidents clearly show that lightning can be a hazard to aircraft. In both cases the thunderstorms were strong. And in both cases the airplane appears to have been clear of the most turbulent part of the storm.

Airplanes are seldom destroyed as a result of lightning strikes, but these two examples clearly show that it does happen and that the phenomenon must be considered a part of the thunderstorm hazard.

Down The Primrose Path

As they approached Minneapolis, the crew of the commuter airline Swearingen Metro talked to another crew about the weather along the route to Lincoln, Nebraska. It would be their next stop after Minneapolis, and the other crew was just returning from Lincoln. Word was that the weather west of the route of flight was dissipating and that there had been no problems.

There was a cold front to the west and a low to the south, down in eastern Kansas. These weren't strong systems, and while there was a low-pressure trough aloft over Nebraska, it wasn't pronounced. Instability was there, though, and the warm southerly flow was moving plenty of moisture into the area. It was a warm midafternoon in June, so thunderstorms weren't unlikely, and the possibility of their occurring was covered in the area forecast. By midmorning the forecasters had added severe thunderstorms for the eastern Dakotas and Nebraska to the menu. Additionally, a series of sigmets covered an area of thunderstorms as it moved toward Omaha. The first sigmet was issued just before noon.

As the flight progressed from Minneapolis toward Lincoln, it operated through a large area of rainshowers and thundershowers, none severe.

Closer to Lincoln there was more significant activity, and at 3:11:23 P.M. the controller handling the flight said, "Wisconsin 965, there's a large area of weather twelve o'clock and about 35 miles, and extends from east of Neola up around Sioux City and down the west side to Fremont. Had one aircraft at flight level 200 went around the area. Another one at 9,000 inbound to Lincoln found his way through it. Don't know how bad it is, or anything else. He said he could pick his way through pretty good, though." The flight replied, "Wilco, according to our radar, it looks okay for us right now, we're 965, we'll continue direct for the time being."

The controller recalled that the entire area in the vicinity of Omaha was covered by lines and H's on his radar, indicative of heavy precipitation. The controller had not received a sigmet that was valid for the area, nor was he advised of severe thunderstorms by supervisory personnel who were aware of the activity. The controller he relieved had told him of weather around Omaha and that some aircraft had been going through it.

This controller contacted the one who would next be handling the flight and told him that flight 965 was "aware of that weather up there but he wants to try to go through it, he said at six." The next controller replied, "Okay, tell him earlier reports indicate light to moderate precipitation and smooth rides." This information was given to the flight at 3:14:31.

After the flight contacted the next controller, who handled the Omaha sector, a couple of other airplanes on the frequency remarked about the weather. One reported heavy precipitation but a smooth ride at 13,000 feet, and the other, at 23,000, reported a good cell about 12 to 20 miles to the right, extending up to about 35,000 feet. The Omaha controller was aware of thunderstorms in the area but hadn't been told of their severity by supervisors. He also recalled seeing some lines and H's on the scope, the computerized radar system's method of depicting thunderstorms.

At 3:23:42 the controller said, "Wisconsin 965, all reports I've got so far at your altitude indicate that you're going to encounter light to moderate precipitation and smooth ride. A Citation reported light to moderate chop and moderate rain, but he was up to about 23,000."

The flight acknowledged, and a few minutes later the controller broadcast a convective sigmet for the area. Shortly after that the flight requested a descent to 8,000. In answer to a query about turbulence, the crew replied, "Light to moderate chop, moderate precip."

Control of the flight was transferred to Omaha approach control at 3:36:06. It would be passing just west of Omaha en route to Lincoln, and the controller stated that the flight was east of the precipitation when he assumed control and that it had "a clear route to Lincoln."

The crew told the controller they were encountering "moderate precip with some lightning strikes to the left and right." The flight descended to 6,000 and then to 4,000. At 3:44:05 the controller asked about turbulence. The reply: "Yes, sir, it's moderate to severe now out here." A passenger later said that even though they were restrained by seat belts, passengers were being thrown around in the cabin. Loose objects were also moving about. Another passenger likened the ride to that on a roller coaster "except you did not know when you were coming down or going back up."

At 3:44:22 the crew requested a still lower altitude

and was given 3,000, the lowest available. Forty seconds later the crew told the controller that they had lost both engines and shortly the crew transmitted: "We gottem both, both of them going."

This was the last communication from the flight. The controller told them that he had a low-altitude alert on the aircraft, but there was no response.

A passenger said that after the aircraft was in increased turbulence for about five minutes, the engine noise level diminished. He looked forward and saw four or five red lights illuminated on the instrument panel. About 15 seconds later, he said, the engine noise increased, but where they had been operating in a "high powerful tone" before the power loss, they were later at a "low tone." He said, "I could tell it was trying to build back up." He recalled activity by both pilots in the cockpit, continued heavy rain and lightning, and electrical discharges from the static wicks on the wings. He didn't recall significant attitude changes, nor did he recall seeing the ground. The other surviving passenger didn't recall seeing the ground before impact either.

There were no ground witnesses. Some heard the airplane go over, and one witness near the site stated that winds were about 100 miles per hour and that there was heavy rain. Winds in the area were measured as high as 70 knots. Broken trees, downed powerlines, and other wind damage in the vicinity of the accident attested to the violence of the storm.

The radar summary chart (Figure 13) for 1:35 clearly depicted the area of activity the flight entered.

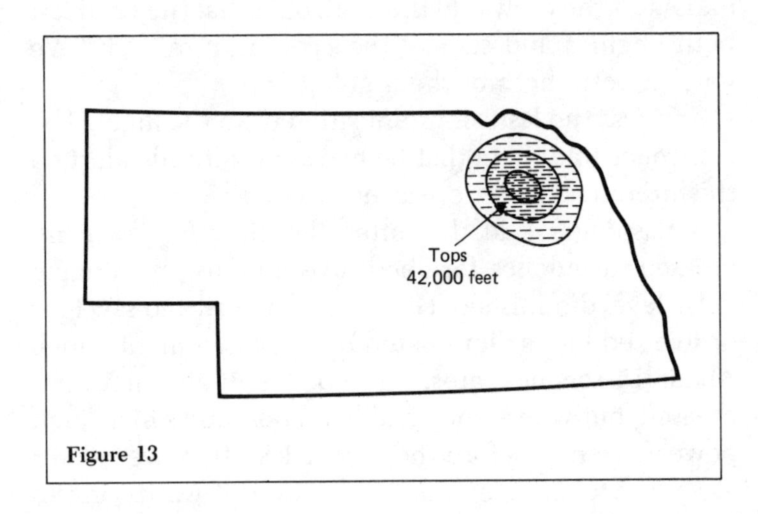

Figure 13

Weather radar units in the area showed extreme echoes in the area, and all the appropriate storm warnings had been issued for area residents.

A videotape of a local TV station's color weather radar system depiction of the storm was used to correlate the storm system and the flight path of the airline, as recorded in air traffic control radar.

The National Weather Service refers to radar echo intensity in six levels, one to six, with one the weakest. Level two precipitation is moderate rain, up to 1.1 inches per hour, with light to moderate turbulence and lightning possible. The radar correlation showed that when the aircraft entered a solid area of level two, there was a level-five (intense—from 4.5 to 7.1 inches of rainfall per hour) cell 40 miles ahead of it. Other radar systems showed a level-six core at this location. Six is about as bad as it can get: rainfall rate in excess of 7.1 inches per hour, severe turbulence, large hail, lightning, and extensive wind gusts.

There was turbulence in the level two area, as reported by the crew, and the surviving passengers' statements indicate that the activity in this area was significant—but nothing like the relatively small but extreme core embedded in the system directly ahead.

The crew had airborne weather radar, but its effectiveness was strongly compromised by attenuation in rainfall. Using ground-based weather radar returns in the area, an engineer testified that the radar in the airplane could have detected the severe storm when about 15 miles away, given the attenuation, but that it wouldn't have given a clue to severity until the aircraft was 9 miles away. Further, he stated that the radar's ability to detect the severe storm would have been even less if the rainfall in the area preceding it exceeded the moderate rate.

The thunderstorm activity was moving from west to east and the flight was southwestbound, so it was on the side that was expected to be the toughest. The presence of the large area of moderate rain and thundershower activity to the east and northeast of a severe thunderstorm isn't, however, what you usually expect to find. That's more the norm when going from west to east. There are exceptions to every rule, though.

The pilots of this aircraft had a lot of things leading them down the primrose path. A collection of optimistic pilot reports and a lack of word from the ground on the extreme nature of the thunderstorms in the area strongly suggested that it wouldn't be too bad. And attenuation caused by general rain prevented their airborne weather radar from presenting a true picture of the way ahead.

7 THUNDERSTORM FLYING

The accidents illustrate the truly sad part. Those pilots didn't make it; some of the airplanes were the same ones we fly, and the pilots were trained and licensed in much the same manner as the rest of us. We've seen how they failed, or at least what led to the thunderstorm-related accidents in which they were involved. There's a lot to learn there. We've also examined the meteorology of the matter, how thunderstorms are forecast, how airplanes are built to withstand turbulence, and who the courts found legally responsible in relation to some thunderstorm accidents. And now it is a cloudy and showery day, with the chance of thunderstorms, or occasional thunderstorms, forecast all along the proposed route of flight. What next?

As seen from the ground, the weather appears benign, flat. But in mind lurks a vision of vertical component—updrafts and downdrafts. There's an uneasy feeling in the gut as the clearance is copied. "Cleared V-190 to Springfield." What is V-190 like? Is it a calm and peaceful blue line on the chart, or does a storm await the little airplane?

The radar summary chart was old and showed only scattered thunderstorms. But the day is more mature now.

The FSS specialist was one of those rare optimists. He said that if any thunderstorms were along the route, center could probably provide vectors. Probably? Does that FSS person realize the stakes in this little game? What if the center doesn't provide vectors? Has that guy ever poked at a thunderstorm in a light airplane? Does he know the feeling of flying into a wall of water and then being depth-charged? It gets lonely when you realize that your hand and mind, and the integrity of the airframe, provide the only buffer between you and eternity. The brain races, the hand trembles, and the airframe creaks in protest when the sky goes dark and the roller coaster ride starts. The cozy and familiar cabin suddenly becomes strange, even hostile.

No pilot who has flown much IFR could deny having thought through such a scenario. And hopefully every pilot *knows* that the key to survival in thunderstorm areas is to work very hard at staying out of thunderstorm cells. Even after the most diligent work, a pilot might occasionally get awfully close to a cell, and the possibility of actual penetration must be considered. But if the avoidance effort is truly and honestly made, and if the pilot is doubly cautious when severe storms are about, the situation should remain manageable, if far from comfortable. The airplane might wind up on the ground at an airport other than the intended destination, but that's okay. The primary objective is to terminate all flights at an airport, any airport, with an undamaged airplane.

KNOW THE SIGNALS

There is hardware to help us deal with thunderstorms. But it's not magic. Back in the fifties, when weather radar was first coming to general aviation airplanes, an old friend and instrument flying mentor got his first weather radar set. Claud Holbert was flying a twin Beech for Winthrop Rockefeller at the time. Mr. Rockefeller liked to operate on schedule. Holbert had been around for a long time, operating from Little Rock, and was wise to the ways of weather. As soon as radar was available for the snout of the Beech, he got it.

Some time later I asked Holbert if the radar parted the waters—if it made a big difference in the way he flew in thunderstorms. The answer was negative. He said that he flew in thunderstorm areas with radar just as he had flown in thunderstorm areas without radar—very carefully, with emphasis on avoidance. The radar provided important additional information, which was good, but it didn't make possible the impossible.

For equipment to be of real value in thunderstorm areas we first have to resolve the important weather question about what might trigger activity. Is there a cold front on the prowl, or a warm front, or a low, or a trough of low pressure? What's going on aloft? Is the wind at the 500 millibar (approximately 18,000 feet) out of the southwest? Is there a trough or low aloft to the west? That's usually the case when severe squall lines develop, but it sure isn't a necessity for thunderstorm development. Is the air stable or unstable? A good clue to possibilities comes from the TV weather on one of the morning network shows. The meteorologists who prepare those maps have access to the convective outlook prepared by the storm forecast center and generally use this to outline areas where thunderstorm activity is possible during the day. Those TV maps have to cover the expected weather developments of the whole day, so they are broad in scope, but they are still valuable. In fact, given the computerized nature of the information that we get from the FSS, a careful consideration of the TV weather map might well be the most important part of a pilot's weather preparation. I've often found this to be a more accurate depiction of the big picture than is available from the FSS.

Another key is flying with an open mind. This is required because nothing ever develops according to plan, especially weather, and it's best to be ready for whatever develops. Even with satellite pictures and computers, the forecasters often miss completely. Then we have to fly with the information gathered as we go along. What is happening *now* is the key to any weather situation.

LOUISIANA BLOW

I saw a classic example of unforecast thunderstorms one day in May. The flight was from Trenton, New Jersey, to Monroe, Louisiana. The weather map featured a low in Illinois with a weak cold front to the southwest and a weak warm front to the east. The forecast for Lexington, Kentucky, our fuel stop, was good. There was only mention of a chance of thunderstorms, with cold-frontal passage predicted for a few hours after our landing. The weather was forecast to be good at Monroe, with not even the chance of thunderstorms.

The flight to Lexington was nice. The warm front was completely benign and was only draping some clouds over the mountains. There was a light southerly breeze at Lex, so the cold front was still to the west.

We were considering a VFR flight from Lexington to Monroe. My son was working on his private license at the time, and if possible I wanted to let him fly this leg with his finger on the map and the transistors on standby. But there were some clouds plus reports of low-level turbulence, so we scrapped that and took off IFR. The FSS specialist said that the weather in Monroe would be good. No problems there.

When we were about halfway, the controller called and in a matter-of-fact manner asked if we knew what was going on at Monroe. I thought things were fine, but that was far from the case. The current weather was 200 overcast, a half-mile visibility with a heavy thunderstorm, hail, and surface wind from the northeast at 13 with gusts to 40. The surface wind had been as high as 60, he said, and a large area of thunderstorms was plaguing Monroe.

We were flying in good conditions and continued on. The activity had moved to the southeast by the time we reached Monroe, but it had been quite something. They had a continuous procession of cells over the airport for about an hour and a half, with occasional hail and very high winds. At one point the wind was strong enough to move a P-3 (a Navy version of the Lockheed Electra) and many of the aircraft on tiedown had moved about in the very strong winds—all this when thunderstorms weren't forecast.

A study of the weather chart for that day shows nothing to indicate an outbreak of severe storms for that area. Monroe was well to the west of a trough aloft, the 18,000-foot wind was out of the northwest, there was no apparent cold pocket of air aloft to create instability, and the surface map showed a very weak cold front to the north of Monroe moving rather slowly southward. The forecaster thus had nothing to prompt any suspicion of thunderstorms; the first mention of them came in the sigmet that was prompted by their actual existence.

There's no mystery about thunderstorms. Something has to happen to get them started. In this case, the nearly stationary front must have actually been in the Monroe area, not to the north, and a low pressure wave must have formed on that front and moved to the east across Monroe. The wave was the trigger mechanism for the activity. That's one plausible explanation; it is supported by the fact that the low-level winds were southerly south of the front and northeasterly north of the front. The moisture supply was plentiful and apparently the heating of the day provided enough instability to support development as the wave got things started. Maybe that's not what happened, but a big part of learning to deal with weather is in taking the situations the forecasters miss and trying to develop an idea of what caused the error. And the very important lesson to learn from this little event is that thunderstorms are where you find them, not where they are forecast to be. Be suspicious; don't trust anyone.

NOT IN THE FORECAST

What if I had been closer to Monroe earlier and had been faced with severe storms that weren't forecast?

Even if you don't have airborne weather detection equipment, there are a lot of clues to thunderstorms. They are, after all, big, spectacular, and hard to hide. Rapidly growing cumulus in an area suggest moisture and instability. A dark and ominous sky is another strong clue. That means heavy rain, and heavy rain is one of the trademarks of the thunderstorm. Lightning might be visible; if not, it could certainly be detected as static on the lowest-frequency band of an ADF receiver. The more continuous the static, the more severe the activity. Finally, once storms get going, someone is going to see them on radar. Any pilot who would fall victim to an unforecast thunderstorm would have to fly right on past a lot of signals. In all the accidents covered earlier in this book, I think it likely that each pilot could have known in advance that he was pulling the tiger's tail.

DO NOT TOUCH

If the decision is made to operate in areas where there are thunderstorms—and this is the key to utility on many flights—then the pilot's job becomes one of moving through the area without getting close to any of the storms. We know the activity is there, and to be successful we must first ascertain that there is a safe path and then find that path.

There are a lot of things to use. Ground-based weather radar information from Flight Service Stations, air traffic control radar, on-board airborne weather radar, or a Stormscope can augment the basic in-flight storm-avoidance gear, human vision. Combine information from as many sources as possible with an understanding of thunderstorms, and with knowledge of the general weather situation confronted, to make good decisions. Any one item of information used in isolation might not work. The more of them used, the better the decision—provided the pilot operates with thunderstorm avoidance as the primary motive. If reaching the destination on time gets first priority, the risk increases dramatically.

Thunderstorms come in lines, clusters, and areas, or individual storms can crop up. They might stand out in awesome splendor, or they might be embedded. Since we know that the greatest turbulence around storms often tends to be in the direction in which they are moving and the direction from which they are feeding, the question becomes "What is the situation?" This needs to be carefully considered before you come close to getting your airplane wet. So does the fact that there are exceptions to all rules-of-thumb.

LINES

Lines of thunderstorms often precede a cold front. We have examined the upper air patterns that contribute to the formation of squall lines and are aware of conditions that can result in quite long lines of thunderstorms. There is continuous lifting along with unstable air, so continuous thunderstorm cell generation takes place as long as all the contributing factors remain unchanged. Lines can also be associated with occluded fronts and troughs, and I suppose that something described as a line could be found in association with a warm front. Relatively short lines can develop in connection with waves on a stationary front or even within what appears to be an air mass.

A fully developed line of thunderstorms offers a most interesting challenge. The storms are likely to be strong, because the situation that results in development of a line is also conducive to strong storm development.

Visual sightings have to be given strong weight when you are dealing with a line of thunderstorms. When you are flying low, in VFR conditions on the east side of a line, the appearance of a roll cloud in front of the line is clearly bad news. Any line of storms (or any individual storm) that generates a visible roll cloud is likely to be strong, probably severe, and the clear message is to be wary. Turbulence generated by the shear between the updraft and downdraft can be extreme well away from the roll cloud or the precipitation areas in the storm. That's not to say that storms without roll clouds are pussycats. It's just that the roll is visible proof of chaotic conditions.

There's a bit of a thunderstorm lore that relates to roll clouds. It probably traces back to World War II, when some missions were deemed important enough to dictate the penetration of lines of thunderstorms without extensive diversions in search of better areas. Indeed, an old military manual on instrument flying states clearly, "Sometime in the course of a pilot's career he will be required to fly through a thunderstorm because of the importance of a military mission." The manual goes on to say, "The least amount of severe turbulence is found at or below 6,000 feet above the average surrounding terrain," and "the roll cloud, which is undeserving of its reputation for containing severe turbulence is found near the 6,000 foot level, and at this altitude turbulence is not severe."

The lad or lass who wrote that had apparently not spent a lot of time around thunderstorms. And perhaps it was this sort of pronouncement that led a generation of pilots to think that the best thunderstorm penetration altitude was at or slightly above the level of the roll cloud. The results of thunderstorm research suggest that turbulence is found at all levels of thunderstorms with the primary difference being in the low-level turbulence being related more to shear between horizontal components (the downdraft fanning out rubbing against the inflow feed of the storm), where at higher levels the shear is more between vertical, updraft, and downdraft components. All the roll cloud tells is where the condensation level is at a point on the slope of the interaction between the downdraft and the inflow of the cloud. It also suggests a very strong interaction between the components. There will be similar rolling action without the cloud at other places along the slope. If the altitude of the roll cloud can be used as anything, perhaps it might be an indicator of where the shear components start becoming more vertical than horizontal.

Cells that make up a line are not always of equal strength all along, and it has been said that when flying low, if you can see through, you can fly through—that is, when looking beneath the general cloud bases, if you can see light on the other side, you can fly through, under cloud, with some success. This theory sometimes works—I've used it more than once—but it must be taken with a grain of salt. Remember, if there is a line, it means that conditions are ripe for cell development all along that line. A set of conditions exists, and the situation isn't likely to change quickly. Individual cells are in a continuous state of generation and dissipation. The fact that you can see through means only that rain is not falling or is not hard. In turn, this suggests no downdraft and thus no fully matured cell. But it might be just about ready to cut loose. There could also be strong updraft activity. And while the maximum turbulence might come from interaction between updraft and downdraft, plenty of turbulence can be found in and around developing cumulus because of the shear between the updraft and air that is generally doing nothing. I've heard pilots say that the coloration of the bottom of the cumulus gives them a clue about flying under a line. That might have some validity—the lighter in color the appearance of the cloud base, the better the deal—but it is no guarantee.

Finally, there can be considerable windshear between areas where there is an active cell with heavy rain and light areas in a line. The outflow from a strong cell will interact with the lack of outflow in an adjacent area.

Any way you slice it, flying through a light spot and under a line of storms based on visual observation carries with it the risk of severe turbulence. And if strong thunderstorms are present in the line, if it is a clearly defined line, and if the gap between strong rainshafts is only a few miles wide, then flying through would be a very good way to get your feathers plucked—and not one at a time.

The strength of an updraft might cause a lot of problems for a light airplane trying to fly under building cumulus, too. Consider that 25 feet per second is an updraft strength that might be found during the development of a garden variety cell. That's 1,500 feet per minute. In calm conditions, will your airplane descend 1,500 feet per minute at maneuvering speed with power off? If it won't, you might not be able to maintain altitude and a proper airspeed, even power-off, in such an updraft. A pilot trying to fly visually under a rapidly building cumulus might find himself exploring the innards of a very unfriendly cloud.

Still flying visually, but high, how about slipping through saddles, or low spots between the storms of a line?

This has worked for a lot of pilots, but when you are evaluating a storm situation visually and from a distance, heights are deceptive. The low spot might be low in relation to other areas in the line, but it might actually be quite high in relation to your altitude. Too, in an explosively developing thunderstorm situation, cloud tops might be spurting upward very rapidly. What looks okay now might be reaching for the stars in 15 minutes. And there is no worse feeling in a thunderstorm area than suddenly realizing that you've climbed as high as you can, the clouds are still higher, you are wandering around castles in the sky, and suddenly you come to a dead end and are enveloped by cumulus clouds. Visual work at higher altitudes can lead to this, especially if observations are through rose-colored glasses.

Adding one more element of information to visual observations of a line of thunderstorms is very helpful.

GROUND RADAR INFORMATION

Ground radar information, from the FSS or air traffic control, can help in defining the depth of a line. This is useful information because, generally speaking, the farther it is through a line, the less severe the activity. If, for example, a line of weather is described as 50 miles through, it might be less severe than one that is 15 miles through. The emphasis is on "*less* severe"; it could still be pretty tough. It's just that in the wider line, energy is being expended over a greater distance. The tightly packed line is probably moving faster, and the faster they move the tougher they tend to be. Again, this is just a bit of general information to be used with other information.

Air traffic controllers will often respond to a request from a pilot for vectors with word that they will take you through the "lightest spot" in a line.

What a controller's radar (or any other radar) sees is

based on the reflectivity of precipitation. What is seen can be affected by the distance from the radar antenna. And our knowledge of what's been learned in thunderstorm research—especially the fact that the areas of greatest turbulence are not necessarily the areas of heaviest rain—should combine to make us very suspect of going through the lightest area in any organized line of thunderstorms. In any case, the width of this "lightest" area would be very important. Remember: 5 miles from garden-variety cells, 20 from severe storms. The controller using narrow-band air traffic control radar sees only two levels of rain, moderate and heavy, and the heavy is identified by H's on the screen. (Virtually all facilities will have only narrow-band equipment as this is published.) So the controller's view is an approximation of the situation as it exists. And a flight through a "light spot" might be very good or it might be very bad. The distance through might be important—if it is a broad line of continuous rain over a lot of miles, the ride might be pretty good. If it's a narrow line, severe turbulence might be present. Whatever the case, the controller clearly does not have information that is completely reliable for the penetration of lines of thunderstorms.

Adding airborne gear—weather radar or a Stormscope or both—brings more information.

RADAR ATTENUATION

The attenuation characteristics of airborne weather radar have to be considered when approaching a line of thunderstorms, or for that matter, when approaching any area of questionable weather.

The dictionary defines "attenuation" as a decrease in energy per unit area of a wave or beam of particles, occurring as the distance from the source increases and caused by absorption or scattering. Point airborne weather radar at an area of rain and you'll get a rather clear picture of the first rain it encounters, although the reflectivity indicated on the scope is influenced by distance. If it looks mean up close, within 10 or 20 miles, it is mean. If it paints at all at 80 or 100 miles on an average general aviation radar, there's probably a lot there and the indicated reflectivity will increase as you get close. At 20 or 30 miles it should give a true picture, at least of the first part of the rain system.

Once you start looking into a rain area, some of the energy from the radar will be scattered and absorbed, or attenuated. I've seen radar attenuated to the point that its effective range is less than 10 miles. That is, the picture on the screen shows the end of the rain 10 miles ahead (or closer), but as you fly along, the distance to the end of the rain doesn't change. It's rather like chasing the mechanical rabbit at the dog track.

Remember, too, that airborne weather radar is painting all rainfall that exceeds a rate of about a half inch an hour as heavy rain. Within a line, the areas of heaviest rain will cause more attenuation. This means that if the range limit of the radar (the end of what you see because of attenuation) is not even, it might be because of a variation in rainfall rate along the line. And what looks like the shortest distance through the line might instead be the heaviest of the rainfall masking Lord only knows what behind the first few miles of heavy rainfall.

So when looking at a line of thunderstorms with radar you might be seeing only half, or a third, or even less of the depth of the line. And you have little or no idea of what's farther along—unless of course you get some information from the controller. If you see 10 miles of weather on the scope and the controller says it's 10 miles through, maybe the picture is a good one. If you see 10 and he says it's 30 through the area, then what you see is only a third of a loaf. Once you fly into the area, rain on the radome will cause the attenuation to become more pronounced. And once in the area of weather, turning around because it gets rough might be worse than continuing straight ahead.

So using a combination of vision, ground radar, and airborne weather radar, the chore is to find a place in the line that appears to offer reasonable passage and to base the decision on information gleaned before the area of weather is penetrated; if the decision is made to go through, most agree that that decision shouldn't be changed.

Don't count on additional information once the area is penetrated. The airborne radar might attenuate so badly that what you get is sketchy at best and can probably be confusing. The controller can't tell you any more once you are in it than he can tell you before you enter. Interestingly, vision might come off best once you are in weather. I've heard more than one pilot profess to have flown toward light spots once in precip, and with some degree of success. This is done on the theory that dark spots are heavier rain and heavier rain is bad. But again, remember that research has shown that the areas of heavy rain are not necessarily the most turbulent. And there's always the moment when all spots look dark.

I had a couple of interesting flights in May that outline some of the dilemmas of lines of weather—and that show the advantages of adding information from a Stormscope to that from airborne and ground radar.

MILD LINE

The first flight was through what I'd have to call an occluded front. The FSS described it as a warm front, and the daily weather map issued after the fact by the National Weather Service showed it as a cold front, so my observation is an average of the two. The characteristics of the situation and the weather behind the front had occlusion written all over them.

There was no mention of thunderstorms before takeoff. I was flying from Trenton, New Jersey, to Lexington, Kentucky. There were some rainshowers, tops probably to 15,000 over the first part of the route. The Stormscope was displaying electrical activity farther west.

As we flew southwestbound down the airway, toward Martinsburg, West Virginia, the Stormscope showed activity to the right. From the rate at which the display of electrical activity moved by to the right, it was fairly close. There was other activity ahead, farther away according to the Stormscope. My weather radar didn't show much of anything until we made a turn to the right to follow the assigned airway after passing over Martinsburg. Then there was an area of rain depicted ahead, light to moderate in appearance. It appeared to be about 10 miles through. In answer to my question, the air traffic controller said there was an area of weather starting in about 20 miles and that it was about 30 miles through. That suggested that my airborne radar was attenuating, not seeing all the weather. It said 10, the controller's said 30. The controller described it as "rain," probably meaning that he was not displaying any H marks, the sign of heavy precipitation, along the line. The Stormscope showed electrical activity to the right, fairly close, and pretty far off to the left. Based on the facts that there was no sigmet, that this was not a clearly defined and strong line, that it was relatively early in the day, that the frontal system was not strong, that the controller reported it as rain, that my radar showed nothing heavy, and that the Stormscope showed no electricity ahead, the decision was to continue.

Once in the area of rain, I found some light to moderate turbulence and some occasional updraft and downdraft activity. As we progressed, I used the airborne weather radar to make small course adjustments, but the radar was attenuating and the adjustments were being made on a rather close-up basis. There were some areas of heavy rain depicted and I tried to avoid all of them. The controller was very close on his definition of the distance through the line, and in total it was a satisfactory passage. To get a true grade on it, I got the readout from the NASA flight recorder that was installed in my airplane at that time. The trace (which records indicated airspeed, g-loading, and altitude) showed the airspeed between 120 and 125 knots most of the way through. That's about maneuvering speed for the weight and was the target airspeed. The g-loading varied from 0 to +2.2. The g-load is 1 in normal flight, so the deviation was 1 in the negative direction and 1.2 in the positive direction. That would be light to perhaps slightly moderate turbulence. Altitude wandered around a bit, plus or minus 200 feet.

A couple of hours later, a sigmet described this area as a line of thunderstorms. What if it had been a solid line when I got there? What if one of the factors that I considered had been negative?

The answers have to be subjective.

If there had been a sigmet for a line of thunderstorms, the chances are the controller would have described it as more than just "rain," the Stormscope would have shown electricity, and my radar would have shown areas of heavy precipitation. In other words, if the situation had been a strong one—one that required a diversion to a better area—there would probably have been more than one negative.

The most likely single negative would have first developed on my airborne weather radar, in the form of heavy rain. I did indeed see some areas of heavy rain as we passed through the line, but I didn't see any in advance. Heavy rain alone does not constitute a thunderstorm. You can have a virtual frog-strangler that will show as heavy on radar without having a thunderstorm. The key is in the transition from areas of light rain to moderate rain to heavy rain, or the gradient. If there is no rain, then heavy rain with only the narrowest band of light or moderate rain between nothing and the downpour, then you are probably looking at a real live thunderstorm. If, though, the rain goes from light to moderate to heavy over a broad area, the chances of its being a thunderstorm are somewhat less likely. So in the case of my Trenton to Lexington flight, the negative factor of heavy rain displayed on my radar in advance might or might not have caused a diversion. It would have depended on the nature of the display, the gradient. The Stormscope indication of nothing ahead was very useful in this context.

When picking a path using airborne radar, staying on the side of the activity with the most gradual gradient is the best way to go. It must always be stressed, however, that the best way carries with it no guarantee of a good ride. If it's a thunderstorm, any part of it can be rough.

It also has to be considered that on this flight I was moving from east to west. The activity was moving toward me. Generally, in connection with cold or occluded fronts, if thunderstorms are embedded in other areas of rain they are more likely to be in the eastern part of the rain area, in the direction toward which the weather is moving. And again, storms usually tend to be meaner in the direction in which they are moving as well as on the side from which the low-level wind is feeding the system. So I could at least imagine that, in the case related, when examining the situation from the east side I was looking at what should be the strongest portion of the activity. Had I been looking at it from the west, trying to look through rain to the leading edge of the system, both my visual evaluation and the information on my airborne weather radar would have been less reliable—the visual because of the masking effect of rain and radar because of attenuation.

I hasten to add that there are exceptions. The accident involving the commuter airline Swearingen Metro was a clear case of an airplane flying southwestbound through a large area of rain and thundershowers before reaching a monster storm. This situation also revealed a flaw in the theory about larger areas of weather containing less severe cells than more tightly packed collections of thunderstorms. That was a slow-moving weather situation with strong instability and a good moisture supply; it clearly illustrated one of the exceptions. In fact, one of the strongest summertime thunderstorms I've ever witnessed from the ground was actually moving toward the west. (And it was in the northern hemisphere, where they normally move east.) The storm had no steering currents in the middle levels, and as it drifted west into a good feed of moisture it built into a big one with gusts to 50 or 60 knots, blowing from the east, as the storm approached.

If you have trouble visualizing the business about one side of a storm being meaner than another, take the time (on the ground) to study the next line of storms that passes your way. The usual scenario is for strong wind to precede heavy rain. Once the rain starts, the wind continues and often shifts around. As the storm passes, the rain usually stops much more gradually than it started and the wind subsides, often before the rain stops. And if the lighting (not lightning) was the same (which it almost never is) the storm would have probably appeared dark as it approached and a lighter shade of gray as it moved away.

SENSITIVITY ADJUSTMENT

Another airborne weather radar item should be considered here.

Some sets have a "gain" control, to reduce the sensitivity of the set. If a line of thunderstorms appears as solid or almost solid heavy rain, turning the gain down will identify areas of rain that are stronger than others. By adjusting the gain, all but the areas of strongest reflectivity can be eliminated. For example, consider a situation where there is a line and the reflectivity value required for a set to display heavy rainfall rate is about 0.5 inch per hour, the norm for depiction of heavy rain. The line contains rainfall rates of from 1 to 3 inches per hour, which means the whole thing would be displayed as heavy rain, equal all the way along. By adjusting the gain control, the pilot could cut out areas where the rain is falling at a rate of 1 or 2 inches per hour and could thus find the areas of very heavy rain as compared with heavy rain. It could be useful if faced with a life-or-death requirement to penetrate the line. And in many situations this could be likened to Russian roulette. It's a bit like taking a vector through the lightest spot from a controller. Lightest in relation to what? And again, all this relates to precipitation, and we know that maximum turbulence and maximum precipitation do not necessarily go together. All factors would have to be considered, with strong emphasis on the fact (yes, fact) that weather radar systems are for avoidance, not penetration. Once the rainfall rate reaches 0.5 inch per hour or thereabouts, and the radar depiction is of heavy rain, the potential for severe turbulence exists if indeed that rain is associated with a thunderstorm. The true value of radar is to give information that is helpful in flying through thunderstorm areas without penetrating thunderstorm cells. The goal should be to keep the airplane dry.

If thunderstorms are really bad and contain tornadic activity or hail or both, there are certain things to watch for on airborne weather radar. A tornado will often make a hook or finger echo—that is, the weather return has a hook or fingerlike appendage in connection with the tornado. If the edges of a storm have a scalloped appearance, that is said to be an indication of a hail-producing storm. And remember that hail can fall well away from precipitation and cloud. The hail itself doesn't show well on radar, so its presence is best judged by the characteristics of the storm.

Where a storm appears to be building or changing shape rapidly, it's probably got something severe going. If we follow the guidelines about avoiding all cells by 5 miles and severe ones by 20 miles, these characteristics are used primarily to decide how far away to stay. Some pilots might make the decision to go closer than 20 miles to a thunderstorm that might be severe, and they might get away with it—especially on a side where the rainfall gradient isn't steep. But it's not good practice, and if doing it once successfully plants a seed that influences future operations, that's very bad. To joust with thunderstorms and form an opinion based on one or two not-so-unpleasant encounters will insure future pounding.

TILT CONTROL

The tilt control on radar is important. This simply tilts the antenna up and down. For best results, set it where the scan is just grazing the ground. Altitude will affect the distance at which the ground first shows as you move the tilt control slowly down.

If the tilt is set too high, the radar might be scanning over the top of storms. Or it might indicate breaks in a line that aren't really there, as it paints only the loftiest chimneys of the storms. If the tilt is set too low, there might be so much ground return, or ground clutter, that the return of a storm is obscured. It must be set correctly. Certainly a weather radar set is a wonderful tool. But it requires understanding and interpretation. And there's still that strong requirement for an understanding of meteorology. Without it, the radar's message might be lost.

BAREFOOT

How would it have worked to fly through the line of weather discussed on pages 228–232 without airborne weather radar and without a Stormscope?

There were positive factors. The controller said there was nothing but rain. The front was apparently not strong at the moment, and there were no current sigmets indicating that the National Severe Storms Forecast Center was tracking significant weather in the area. I would have probably flown through. If the controller had told of heavy rain in the line, I probably wouldn't have flown through; instead, I'd have asked if there wasn't a route free of heavy precip. The ADF might have played a role here, too; if there was no static with the ADF tuned to a low band, I'd have been optimistic. If there had been continuous static, my goal would have been to look for a better westbound path.

It might be argued that flying a path chosen with the help of a controller necessarily involves sweaty palms. But that should not be true. If a pilot will study the general weather situation, he will have enough knowledge to avoid the really bad stuff. He'll probably detour a lot, but some of the best flying time is that spent giving a wide berth to thunderstorms. It must be added that localized heavy showers with updrafts, downdrafts, and turbulence can escape the H designation on the controller's scope. Every pilot who has flown much while using information from the air traffic controller's radar for weather avoidance has experienced wet and bumpy moments. But for that matter, so has every pilot who has flown much with weather radar or a Stormscope.

MIDWESTERN LINE

My other May experience with a line of storms came between Wichita and Indianapolis.

I was awakened by thunder, which always outlines the chore of the morning if you are about to fly away to the east.

The call to the FSS revealed what was obvious: there were strong thunderstorms in the Wichita area, moving to the east. Additionally, the radar summary chart depicted activity along my route. It extended as far as central Missouri. The synopsis showed a low to the southwest of Wichita, with a warm front extending to the east. The upper level trough was to the west. The situation was not quite ripe, but it was clearly one that might develop severe stuff later on.

This flight was also made with both radar and a Stormscope. Before I took off, the Stormscope indicated an electricity-free path to the northeast. The big storm that had substituted for my alarm clock earlier was to the east and southeast and was well outlined by the Stormscope.

The climb to 19,000 feet was relatively smooth and with a good tailwind. There was nothing painting on my radar, and the Stormscope showed activity to the right. But when I was about 100 miles northeast of Wichita the controller called and told me that there was a line of weather about 20 miles ahead. I couldn't get a clear picture of anything on my radar, and the Stormscope showed all activity to be to the right. Just to make sure, I turned a little to the left and gave extra margin to the electrical activity on the Stormscope. The line of weather described by the controller was there, the precipitation in it was snow, heavy at times, and I couldn't get a good paint on my radar because snow doesn't have the reflectivity of rain. The snow would indeed change to rain at lower altitudes, which could be detected by proper use of the radar antenna tilt, but that's subjected to limitations because at some point the radar returns will become mixed with ground clutter.

At any rate, my ride through the controller's "line of weather" was good. It was a broad area—about 35 or 40 miles—and there were only some light bumps in there. However, after I got into clear air to the east I had quite a rough ride on to Indianapolis at 19,000 feet. Other aircraft were complaining of turbulence at all levels.

Something else that happened illustrated one of the problems forecasters have in pinpointing where strong thunderstorms are likely to develop. The forecast wind I got for the 18,000-foot level suggested a relatively strong and consistent flow from 260 or 270 degrees; the 500-millibar chart on the official daily weather map published after the fact by the National Weather Service agreed with this forecast. But there was a wiggle in the pattern at 18,000 feet that had an effect on the development of thunderstorms that day, and that was finally shown on the 500-millibar chart the next day. It also showed on the groundspeed readout of my DME—24 hours before the Weather Service got it to a map. The groundspeed was good in the climb and reflected a strong tailwind up until the time I reached the area of weather. Then the groundspeed started deteriorating and was finally down to a value about equal to the true airspeed—despite the forecast of a uniform strong tailwind. It was turbulent on to Indianapolis as the groundspeed slowly started moving back up, reflecting a renewed tailwind.

Figure 14 shows the 500-millibar charts for May 13 and May 14. The 13th was the day I flew the flight, but the map for the 14th more nearly reflects the situation encountered. The actual upper-wind measurements

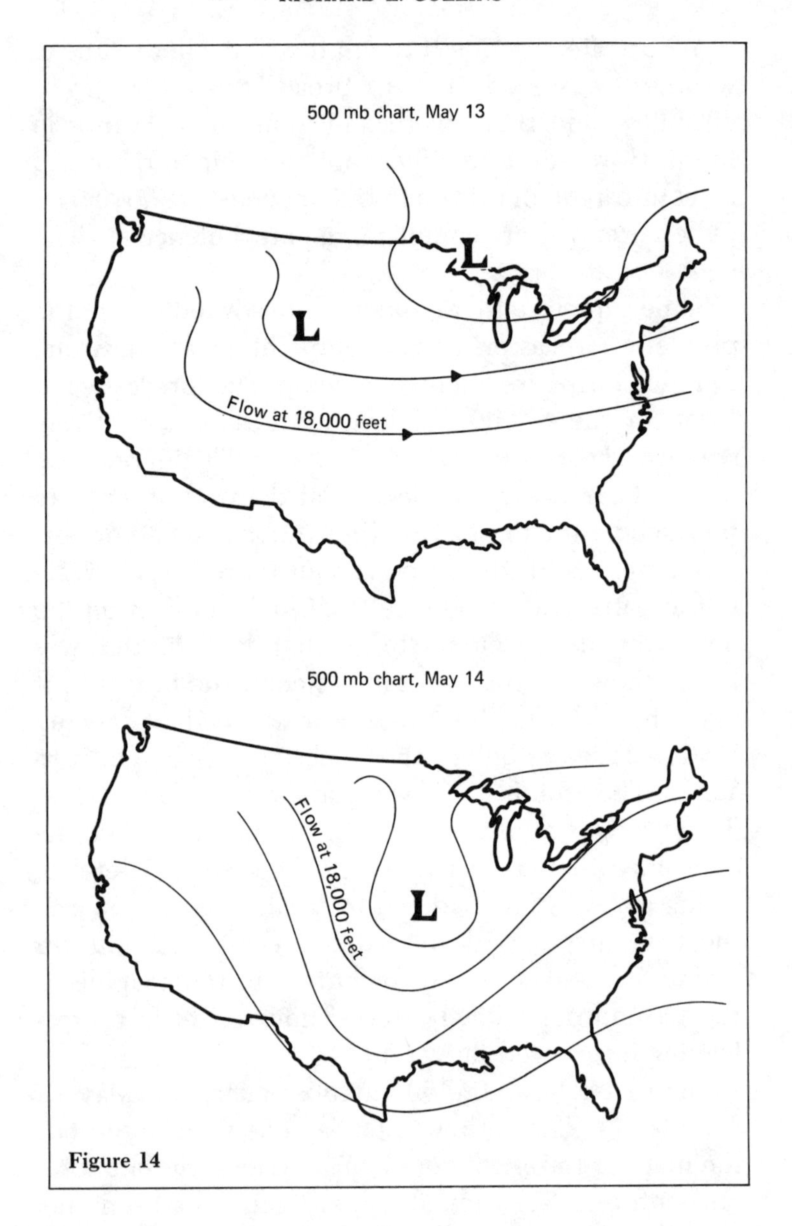
500 mb chart, May 13
L
L
Flow at 18,000 feet
500 mb chart, May 14
Flow at 18,000 feet
L

Figure 14

are old when the forecasters get them, and as I've said before, there are often gross inaccuracies both in the wind forecasts and the information that forecasters use to predict thunderstorms. All the more reason for pilots to develop an understanding of weather based on what is there as opposed to what is forecast and, when dealing with thunderstorms, to pay attention to the sigmets which describe situations that have developed.

Neither of these so-called "lines of weather" was a squall line, or an actual line of thunderstorms. Instead, they were lines of rainshowers with some embedded thunderstorms—an entirely manageable situation as long as you find the good path through. A strong squall line is an entirely different matter and dictates an entirely different approach. Sure, airline, military, and general aviation pilots have penetrated squall lines and survived. But the risk is very high and, given a strong enough squall line and presence at the wrong place at the wrong time, the turbulence in and ahead of a squall line can break even a heavy airplane.

HONEST, THE WORLD IS ROUND

Radar pictures of squall lines are usually available for study before flying, and there's something to consider when looking at radar pictures of lines of thunderstorms. Weather radar antennae rotate all the way around, making a circular picture. Lines of thunderstorms tend to be straight, so the range from the radar antenna to the various points of the line is not equal,

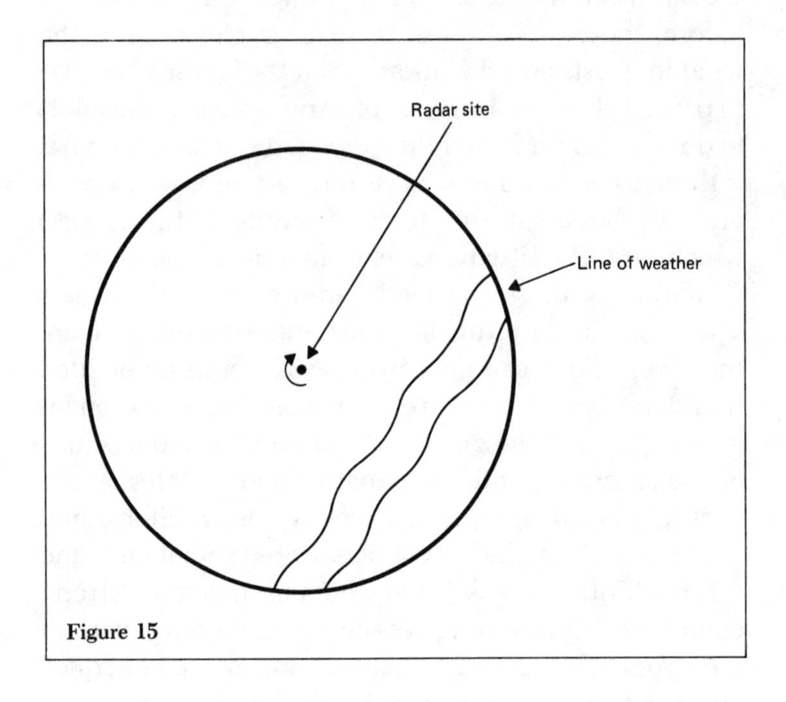

Figure 15

as illustrated in Figure 15. The picture will thus suggest that the strongest part of the line is where it is closest, with activity appearing weaker where the line is farther away from the antenna. When the line is a good distance away from the radar location, it might even appear that there is an end to the line, because the radar is actually looking over the top of the storms at extreme range.

I learned a good lesson on this when flying a Skyhawk from Fort Worth to Little Rock one day.

A line of weather had been developing, and all day I anticipated an interesting trip home.

The Fort Worth FSS had a facsimile picture of the local weather radar. That's not the best quality reproduction, but it gave the general picture. There was a line of weather around Texarkana, 170 miles to the east. The line was oriented northeast–southwest. The FSS specialist suggested that I should be able to go around either the north or south end, as the line didn't appear to be all that long.

En route and in contact with the air traffic controllers, it was a different story. If I remember correctly, a controller told me that the line extended from Chicago to the Gulf of Mexico. He could offer no "light spots" or anything else.

When we overtook the line it was about 20 miles east of Texarkana and very interesting in appearance. Approaching from the west side it didn't have an ominously black appearance. There was gray, and the showers appeared to be scattered. I flew into the shower area far enough to see that there was nothing but a wall of darker gray at the end of each gap in the showers on the west side of the line. It was a clear mandate to retreat and have dinner in Texarkana.

There were some other people there waiting to go on and I heard an interesting comment from a passenger on one turboprop. The pilot of the airplane had examined the line (from the west side) visually and with his airborne weather radar and had made the decision to land and wait awhile before continuing eastbound. The passenger was unhappy about the delay. He was also a pilot, his company operated corporate aircraft, and he said that based on the radar picture he would have flown on through. It was probably best that he was with a pilot who flew using all the information at hand, though, because the situation was one of those classics where attenuation caused by rain and rainshowers to the west of the really strong stuff would have masked the true character of the line of storms. This day, the forecasts of severe storms, the length of the line, and the fact that air traffic control radar was indicating heavy rain all along the line, dictated that any optimistic picture on airborne weather radar be disregarded. Certainly from the other direction, where the strong cells would not have been obscured by attenuation, the line would have appeared impenetrable on airborne weather radar.

CLUSTERS AND AREAS

Clusters of thunderstorms might be found along a front where conditions aren't uniformly conducive to storm development. But they are more likely to be found in areas ahead of a low-pressure center and where moisture and instability are present over a large area as opposed to along a front. Given pilot determination to avoid rather than penetrate, clusters shouldn't offer great problems. Remember, most storms have more than one cell, so when we are talking clusters of thunderstorms the reference might be to a large area, 50 or more miles across, or to a smaller area that might from a distance appear to be one thunderstorm.

Again, the key is in identifying and avoiding all cells.

I was flying my Skyhawk from Little Rock to Wichita in the fall of the year. In the vicinity of Fort Smith, Arkansas—about a third of the way to Wichita—I started moving into an area of building cumulus. I was IFR at 10,000 feet—high for a Skyhawk—and the clouds soon built and enveloped the airplane with bumpy wetness. The controller told me there were areas of rain ahead and I asked for vectors around. There was a relatively strong low-pressure area to the west of my position, and the storms were developing ahead of that low.

The controller suggested that I deviate to the north, and flying in that direction I soon got the Hawk back out of the building cu. It didn't look good in the direction that I wanted to fly, but by keeping the nose pointed at conditions that were friendly in appearance, I was at least not getting any farther away from my destination. And hopefully I was improving the quality of the weather between my airplane and the destination.

After some more miles of flying, the cloud tops lowered and I could clearly see an area of strong thunderstorms off to my left. And some of my thoughts clearly illustrated how careful we must be when navigating in areas of thunderstorms using visual impressions as well as information from ground radar or other sources.

I was still on a northerly heading, and my destination was by this time west of my position. I had thus far avoided clusters of thunderstorms and the tops of building cumulus simply by staying east of the area of strong lifting. At some point I had to turn to the west. I saw one area that looked good; the heading would have been southwesterly. In response to my question, the controller said that it appeared clear on his radar, but that it wouldn't get me any closer to Wichita and would in fact lead to a position where there would be even more weather between me and Wichita.

In answer to my question about another potential turn he said he didn't see any opening there. And on closer study I realized that what I thought was an area of blue sky was actually the green appearance of very heavy rain with just the right lighting effect. I don't think anyone would actually fly into such a thing based on a visual observation, but I sure would have flown toward this one if the controller's doubt had not suggested that it was other than blue sky.

Clearly, I was dealing with numerous clusters of thunderstorms in an area where conditions were ripe. And the only way I was going to get to Wichita was by reaching a position where a gap between clusters would let me fly through. I finally got to this point when northeast of Wichita. The controller said that a straight shot looked good, and I had a smooth ride on in. I couldn't calculate the length of the detour, but the trip probably took only an additional 45 or 50 minutes because of the deviations around weather. It was only a matter of searching until I found a path that I felt sure was neither primrose nor thorny.

In that case, there was no doubt that real live thunderstorms were on the prowl. The synopsis (a strong low to the west), the moisture supply (a strong southerly flow from over the Gulf), the visual appearance of explosively building cumulus, the view of real storms as I flew closer to the low, the controller's description, and the sigmets describing areas of thunderstorms, all gave clear definition to the situation. Such is not always the case.

LANGUISHING FRONT

I was flying from the Cleveland area to New Jersey. There was a dead or dying front between the two points, one that had defied the forecasters for several days. They had kept insisting that it would move off the east coast; the front chose to languish inland. From my flights in the area during these days it was apparent there was little action aloft—almost all flights up to as high as 17,000 feet were with very little wind—and the surface conditions didn't suggest that anything was happening. The winds at the surface were light northeasterly behind the front and southerly ahead of it. Some waves were forming and moving along the front, but nothing big seemed to get going.

Still, the forecasters clung tenaciously all week to "occasional" thunderstorms.

As I moved toward the east on this trip, there were some buildups of thunderstorm size visible to the south. The controller said that there was activity to the south, verifying this, but he said he didn't see anything along my route toward Johnstown, Pennsylvania. I was flying at 17,000 feet in a P210 and made a few detours based on visual sightings. One cloud was particularly interesting. It was big (tops probably to 25,000 feet), bubbly, and certainly worth avoiding. Flying into it would have resulted in a thorough pummeling. Yet it reflected only as light rain on my radar and it never made a mark on the Stormscope, even with the 200-mile, high sensitivity selected. The controller never saw it. It was a good illustration of how human vision should be allowed to override electronic observations. Later, as I flew eastbound, the Stormscope started picking up a lot of electrical activity behind and I assumed that this cumulus had ripened into a cumulonimbus.

Closer to home, the New York center controller was telling pilots of weather and approving deviations. This was significant to me, because the controllers at New York center seldom ever mention weather. They usually leave that little matter entirely to pilots.

As I flew closer, the controller described the weather as being around Harrisburg, to the south of Harrisburg, and around Lancaster. He said that a deviation to the north would be okay, but the picture on my radar looked better on course, over Lancaster. The controller said it was okay with him if I wanted to go that way; some air carriers had gone straight through without a bobble and others had deviated around cells in the vicinity of Lancaster. The Stormscope showed no electrical activity ahead (there was some to the south) and I used this as an indication that there was no tough stuff in the area.

When I flew into the first of the rain, there was clearly some updraft activity. I had to extend the landing gear to maintain a reasonable power setting as well as the assigned altitude. The rain was of the very showery variety—a good shot of heavy, followed by lighter—and the radar reflected this. However, there were only a few areas of heavy rain, and these didn't have the appearance of being thundershowers. I flew through with only minor heading adjustments and light to very occasionally moderate turbulence. The view was a relatively dark shade of gray, and at times a billowy cumulus would float by, embedded in the rain and other clouds. I guess the few bumps I hit were related to these clouds. The area of rain, which was 70 miles through, dissipated in place about four hours later. It had probably resulted from the formation of a wave on the front, a wave of only modest ambition.

Again, it was an example of the importance of knowing the situation and combining all available information.

A TROPICAL LOW

A most enlightening encounter with a large area of clustered thunderstorms came in June. The weather situation—a strong low of almost tropical-storm strength—was unusual enough to set some rainfall records in Arkansas, Mississippi, and western Tennessee. The low was in Louisiana on the day of the flight. It had been strengthening for several days, with rather continuous periods of heavy rain over Arkansas. This day the FSS briefer didn't mention the low at all, but the morning TV weather map showed it to be in Louisiana, and the prediction was for thunderstorms and heavy rain to the north and northeast of the low. My flight was from Trenton, New Jersey, to Lexington, Kentucky, and then on to Little Rock, Arkansas.

There was a little shower activity on the route to Lexington but nothing of consequence. A check of the weather at Lexington revealed widespread activity along the way to Little Rock. The flight, in a P210 equipped with both weather radar and a Stormscope, was to be one of the best flights I've had for the purpose of comparing the two weather avoidance systems. It also clearly illustrated some advantages of air traffic control radar for storm avoidance.

I hadn't flown far to the southwest of Lexington before weather became a factor. Visual observation won on the first activity of this leg: the dark gray was clearly visible from quite a distance. A layer of clouds at about 15,000 feet obscured the tops, but I thought they were likely to be high. The radar painted only light to moderate rain, and the Stormscope depicted no electricity. The controller had a good paint on this weather and suggested that I go to the east and south of it. Passing rather close, I looked it over carefully and decided that it was dissipating, which explained the lack of electrical activity. I'd say that it was still worth avoiding, as there were some strong cumulus building amidst the remains of what had probably been a cumulonimbus an hour earlier. These might well have later developed into active cells in their own right.

Once by this, I had a discussion with the controller about the rest of the route to Little Rock. He strongly suggested that instead of flying southwest I head straight west toward Paducah, Kentucky, and Malden, Missouri, and then on into Little Rock. There was, he said, considerable activity down over Memphis, the way I'd normally have gone. Sigmets were being issued on this area, and in addition there was a flash flood watch. The controller added that there would be some weather on the suggested route but that it appeared scattered.

This bit of advice illustrated the great value of current weather information from ground-based sources. They can see a bigger picture and can help plan a route that will give maximum advantage to the on-board weather avoidance devices. Left on my own, I'd have flown into an area of closely spaced activity where deviating would have been a tedious process.

After going around that first little bit of weather, I was operating in an area of building cumulus that were becoming cumulonimbus. Light and moderate rain was painting on my radar, tops were about 20,000 or 25,000 feet and going higher, but neither the Stormscope nor the controller's radar was painting a thing. These cu didn't yet qualify as thunderstorms, but they were still worth avoiding.

A little farther on, strong activity was evident on the Stormscope and I was reminded of a situation in which the display of this device has to be carefully interpreted.

There was an active cell about 40 miles ahead, from twelve o'clock to one o'clock. Another cell was out at about 65 miles, from eleven o'clock to twelve o'clock. On the Stormscope the return from the two merged and it appeared as one big cell, from eleven to one o'clock, even though there was plenty of room to zig south a little around the first one and then zag north a little around the second one. As the range narrowed, of course, the Stormscope presented a truer picture.

The controller continued to describe the best route as being just north of Malden, Missouri, and then toward Little Rock. Both the Stormscope and the weather radar strongly agreed with this. I felt that I was getting an accurate depiction from both systems as I flew carefully toward the turning point.

It was in back of the area of weather that my information sources started disagreeing.

After I turned toward the southwest, the controller said I could go straight to Little Rock with only minor deviations. The Stormscope showed no electrical activity ahead. But the radar in the airplane showed quite a bit of activity, with some heavy rain depicted. The areas were relatively small and they lacked the very sharp gradients of a strong thunderstorm, but avoidance appeared advisable. I wandered around, changing heading by not more than 30 degrees to avoid the areas of rainfall that were associated with the depiction of heavy rain. The ride was smooth. The airplane was in cloud most of the time, but occasionally I'd be in visual conditions and could see the flank of a billowing cumulus off to one side, a reminder that my wandering was worth while. Another reminder came from an airplane that had also taken the controller's routing around the weather and was now southwestbound. This pilot, apparently flying without radar, was flying the airway without deviating. He occasionally reported heavy rain and moderate turbulence. There was nothing dangerous in those showers. They weren't painting as heavy on the controller's radar and weren't making a mark on the Stormscope, but the ride was far better outside than inside. The level of rain and turbulence in those showers was probably strong enough to make the pilot of a light airplane feel that he had at least flirted with a thunderstorm.

On this flight the information from any one of my

three sources would have outlined a safe path around the rough stuff. But the radar in my airplane did the best job of defining the most comfortable path.

INDIVIDUAL AIR MASS STORMS

Thunderstorms that develop within an air mass—usually in the summertime—are often considered no problem. They can be easy to see, they cover a relatively small area, and they don't last too long. But pilots can find trouble in and around them for several reasons—especially when a little disturbance develops within the air mass, fools the forecaster, and allows development of more than an isolated cell.

In areas where haze is prevalent, such as along the east coast in the summertime, afternoon thunderstorms are embedded in the haze. A pilot flying low-level VFR or IFR without weather mapping equipment becomes aware of them only by the darkening sky ahead. The storm might not assert itself visually until only a mile or so away. Even then, it is only a darker coloration in the murky view. Needless to say, the nose of the airplane should always be pointed away from such areas.

A clue to thunderstorms embedded in haze can come from cumulus clouds. If they are few and far between and don't show much vertical development, then the instability isn't strong and the moisture supply isn't plentiful. On the other hand, if it is a hazy day with a lot of building cumulus embedded in the haze, there is moisture, and there is at least some instability. Thunderstorm activity should be suspected.

If there are any hills or mountains around, thunderstorms can form over the hills (because of lifting generated by wind flow up the slopes) and can develop all along a ridge. Any pilot who has flown across the Appalachian Mountains in the summertime has tales to tell about being bounced around and wet down even when the controller doesn't report heavy weather and when there is no feature on the weather map to indicate thunderstorm activity. An easterly flow along the coastal plain causes lifting when it reaches the mountains, and this can and does result in the development of thundershowers over the ridges. Because these showers often don't develop to a height that becomes an operational consideration for jets, and because they are seldom severe, this activity is for a fact often unnoticed by observers and forecasters. But it can become uncomfortable for general aviation pilots operating at the lower altitudes. Rocky Mountain ridge lines often have quite an array of thunderstorms when there is a supply of moisture.

One thing should be stressed when considering air mass thunderstorms. Just because they don't cover a large area, and just because they are a part of the afternoon scene in many parts of the country, don't consider them weak. The strong afternoon thunderstorm is just as worthy of avoidance as a squall line.

"FLYING" THUNDERSTORMS

Most of us put total effort on thunderstorm avoidance. Some don't, and as we watch them fly it's only natural to wonder how long it'll be before they get caught.

I was flying along in the southeast one day, in my

Cherokee Six, and saw thunderstorm activity well ahead. I asked the controller about it and he said that if I'd turn right 20 degrees and fly that heading for the next 75 miles I'd pass north of what he was painting on radar. Fine.

As I was flying north of the activity, a pilot flying a Cessna Skymaster came up on the frequency, IFR and bound in the same direction as I. The controller told him about the weather ahead and offered a vector. The pilot asked how far out of the way he'd have to go; the controller replied that it would probably take about 20 miles of extra flying. The pilot said that he'd fly on course. The activity was what would probably be classified as thundershower rather than thunderstorm, with cloud tops not a lot over 25,000. The pilot flew through without incident and without comment. Maybe he just likes turbulence.

In another case, a friend of mine, an FAA control tower chief, rather excitedly related an incident that had taken place at his airport. An airline jet had started the takeoff roll as a strong thunderstorm was moving onto the airport, over the departure end of the runway. Right after the aircraft started rolling, the tower controller called the flight with word on the surface wind. It was gusting to over 50 knots. The aircraft continued its roll, rotated normally, and then flew into the wall of water produced by the storm. The rain was so heavy that the controller's radar (of the old broad-band variety) would not even paint the aircraft's transponder.

The controller tried to call the flight, but there was no answer. This caused strong suspicion that a thunderstorm had won an argument with an overambitious airline captain. But a few minutes later the transponder return emerged from the back of the weather and the flight called. The crew apologized for not answering previous calls from the controller, but they had been "ahhh, kinda busy there for a minute."

IT'S IN WRITING

The fact that flying into a thunderstorm isn't considered automatic suicide is found in writing, too. That old military manual I quoted earlier (p. 221) put it thusly:

> Thunderstorms do contain severe turbulence and should be avoided when possible; however, when flying through most thunderstorms it is easy to control the aircraft and maintain a reasonably constant altitude and heading. Any pilot who can fly well on basic instruments and does not become excited by lightning flashes and turbulence can safely fly through thunderstorms with any tactical-type military aircraft.

In older books aimed at (or written by) airline pilots when they were flying airplanes of about the same speed and performance characteristics as today's general aviation piston airplanes, there is often a description of the procedures to use when flying in a thunderstorm. Even relatively recent government publications aimed at general aviation audiences have contained information on techniques to be used when flying through thunderstorms.

This is as it should be. It is probably certain that any

pilot who extracts utility from an airplane will some day graze, if not penetrate, an area of moderate and perhaps severe turbulence when flying on instruments in an area of thunderstorms. So to do anything other than explore all the possibilities can leave the pilot playing with all the chips on the table and less than a full deck.

Primary factors that bode well for a pilot's making it through an inadvertent cell encounter are an understanding of meteorology and a slavish dedication to the avoidance of thunderstorms. These factors will hopefully combine with good judgment to result in any brush being with activity that isn't of a severe nature. In other words, if you try hard to avoid them all, any penetration should be with a cell that isn't severe. Given that, the airplane's controllability and structure should be up to the task. This leaves only the requirement that the pilot have a cool head and good instrument-flying capability.

I can't go along with the old Air Force book's statement that it is "easy to control the aircraft" in most thunderstorms. It can be pretty darn trying. But there is a big key to any difficult flying situation found in that book's admonition not to become excited.

There is no question but that the tension level rises when flying in areas of thunderstorms. If the pilot has his wits about him, a lot of concentration goes into avoidance, into seeking the best way through the area. Those who do it with the least tension are probably the pilots who study the situation carefully, make conservative decisions and, once a path is chosen, tighten the belt, loosen the grip on the wheel, and fly with a dedication to maintaining a wings-level attitude through whatever comes up. Remember, too, that once in the difficult situation, the primary requirement is to *fly*. Some pilots seek aid and comfort in the microphone after penetrating an area of turbulence. This can only divert attention. If you were in contact with the controller before entering the weather, you probably got whatever information he had. Further conversation won't add much to that, and it's a cinch that the controller can't help the pilot fly the airplane.

BE GENTLE

There are fine points to handling the airplane in turbulence. In even the wildest situation the pilot's control inputs can have a lot to do with the quality of the ride and the stresses imposed on the airframe. In most cases the smooth control input that is applied after an instant of thought is probably the best. Rough and jerky inputs are bad, and failing to evaluate the true need for an input could result in doing the wrong thing at the wrong time. If, for example, the airplane rolled rapidly right and you made a snap judgment after misinterpreting the instruments, you might move the wheel in the wrong direction. And in turbulence it's certainly not impossible to momentarily misread an instrument. Or the airplane might roll right, and the pilot might honk in a massive amount of left aileron just as the airplane moves into the next bit of shear, which rolls the airplane to the left. This combines with the control input that's already there to result in quite a rolling movement to the left.

Whenever a pilot is flying in turbulence, it's quite

easy to evaluate the contribution of pilot inputs to the bad ride. Just hold the wheel still for a moment. In most cases that will result in a better ride. But it takes a lot of self-control to resist reacting with control input to every little jiggle of turbulence.

FREE ADVICE

There's plenty of advice on how to operate in and around thunderstorms.

The words in the government publication *Aviation Weather* prompt good thought on the subject. The section on actual thunderstorm penetration includes the following:

Avoid altitudes from the freezing level upward to −10 degrees C, roughly 5,000 feet above the freezing level. The publication also claims that the "softest" altitude in a thunderstorm is usually between 4,000 and 6,000 feet. This advice is refuted by recent thunderstorm research flying in which turbulence was found not to vary much vertically through the storm. However, generations of pilots have stuck to the "lower is better" theory, and it shouldn't be completely discounted. The shear at lower levels is a product of the interaction between horizontal forces more than between vertical forces, as at higher levels. So there is a difference, and some feel that light airplanes do better with the horizontal interaction. Too, in a warm frontal area the bases of the storms might be high, and flight below the bases would be better than flight above the bases. And of course the advice to avoid altitudes from the freezing level to −10°C will help avoid the misery that might come from trying to fly a well-iced airplane in a thunderstorm after a lightning strike.

Change airspeed to the manufacturer's recommended speed for turbulent air penetration. If there's no separate turbulent air penetration speed in the pilot's operating handbook, maneuvering speed would be considered the best speed for turbulence. Remember that this speed decreases at weights below gross, as is covered in most handbooks. This best airspeed for turbulence should be used whenever turbulence is suspected (as opposed to expected). In other words, don't wait until you are in the stuff to slow down.

Keep constant power settings in the storm. This is a rather broad statement. Certainly it is applicable to heavy airplanes that are somewhat less vulnerable to rapid accelerations and decelerations than light airplanes. However, where power response is rapid, as in a piston-powered light airplane, changing power settings seems less of a no-no. If the altitude and airspeed are both showing an increasing trend, for example, a smooth reduction in power while gently holding the airplane in a level attitude might help more than it hurts. Likewise, if the ship is sinking fast and the airspeed is decaying, the application of power shouldn't be ruled out. Too, if things get really bad, then the rules must change. In most cases where a general aviation pilot loses control of the airplane in a thunderstorm and eventually fails the airframe, power is not reduced (and the gear is not extended on a retractable). Perhaps this is a result of panic, or perhaps it is a result of feeling that nothing should be changed. However, if control of the airplane is in doubt, extending the landing gear means the airplane will build speed less rapidly in case of a loss of control. Likewise, responding to rapid increase in airspeed with a reduction of power will result in the best possible configuration in case of an upset.

Fly in a straight and level attitude. At least try to fly in a straight and level attitude. And be gentle about it. The turbulence will put enough loads on the airframe.

If the autopilot is used, the altitude hold mode should be in the off position. This directly relates to the previous admonition about the straight and level attitude. If altitude hold is on and the airplane enters an updraft, the autopilot would put the airplane in a nose-down attitude to try to maintain altitude. It'll be vice versa in a downdraft. It's okay to leave the rest of the autopilot on if it does a smooth job of flying the airplane in turbulence. If it's jerky, or if it seems to get behind, perhaps the pilot would do a better job. Whatever the case, when the autopilot is flying in turbulence, the pilot should be ready to help.

Pick a heading that will take you through the storm in minimum time and hold it. This relates to the drill about not turning around once in a thunderstorm. And it is probably true most of the time. However, if penetrating weather from the back (usually the west side in the northern hemisphere), where it's likely to get worse before it gets better, perhaps this point should be qualified. When penetrating a line of weather from the west, with the radar attenuating, if stronger cells appear ahead and the turbulence is still manageable, some course alteration (up to and including a 180) might be considered *if* the turn can be completed before the area of stronger cells is encountered. On the other hand, if the pilot uses good judgment this situation will not arise. There are some good reasons for not turning around once the going gets rough. For one thing, whatever made you want to turn around would have to be flown back through after the 180-degree turn. If it was *that* bad, the thinking goes, maybe the worst part was already behind. For another, there is probably some relationship between attempted hasty turns out of turbulence and loss of control. When the airplane is in a wings-level attitude, it is as far from trouble as possible. If turbulence rolls it into a 30-degree bank, the airplane is still not in a tenuous position. Let the pilot honk it into a 30-degree bank to get out of a wildly turbulent situation and the bump that rolls it 30 degrees might put the airplane into a 60-degree bank—pretty close to a lateral upset.

There are other items of advice on thunderstorm flying that are worth a moment.

The tightness of safety belts is often emphasized. This is good. It is harder to see and concentrate on the instruments when you are bouncing around in the seat, and you could bump your head if the belt is loose. Being moved about also results in a general feeling of instability. How tight? As tight as you can get it. If the airplane has a shoulder harness, it should be tight too. Unfortunately, most inertia-reel harness systems don't have provisions for manual locking, so they are of no use in turbulence.

Turn on the Pitot heat and other appropriate engine and propeller heat. This is to keep the airspeed indicator working and the engine running smoothly in case of icing—both worthy causes.

If it's nighttime, turn the cockpit lights to maximum intensity to minimize blindness caused by lightning flashes.

Before reaching a storm area check the cockpit for loose items that could hurt if thrown about the cabin by turbulence.

Don't look outside. Doing so could accentuate blindness caused by lightning. More important, you can fly instruments only by looking at the instruments; time spent looking outside can only be considered distracting. If you are trying to fly toward light spots in an area of heavy rain, don't let attention to this divert attention from keeping the wings level.

Remember, too, the definitions of turbulence both when evaluating pilot reports and when making your own.

Light turbulence causes slight changes in attitude and altitude.

In moderate turbulence, altitude, attitude, and airspeed changes occur. Strains against belts are felt and loose objects move about, but the airplane is still relatively easy to control.

Severe turbulence results in large and abrupt changes in altitude, attitude, and airspeed. Loose objects are thrown around, and there's considerable straining against belts. The airplane might be (or seem) momentarily out of control.

Extreme turbulence is worse than severe turbulence.

If anything is clear as we read thunderstorm advice,

it is that there are exceptions to all rules. Blanket statements must always be qualified.

PANIC

Where a thunderstorm accident is related to a loss of aircraft control, the possibility of pilot panic must be explored. Turbulence within a cell might be easily survivable from an aircraft structure and controllability standpoint, but it might frighten the pilot. Light airplanes are relatively small in relation to the shear and the eddies. They can thus be affected in both the roll and the pitch axis. And while pitch might be the least serious of the two (because airplanes are inherently stable in pitch and will by themselves attempt to maintain the trim airspeed), the turbulence-induced deviations in pitch attitude are probably the most disconcerting to the pilot. When severe turbulence is defined as that where unsecured objects are tossed about the cabin, it is often the pitching moment of the aircraft that does the tossing. Pitching can come from the sudden entry into up or downdrafts or encounters with strong shear. It can also come from the rolling and tumbling of air in the area between updrafts and downdrafts, and between the inflow and outflow associated with the storm. Once when flying my Cherokee Six at what I thought was a safe distance from a cell, I entered an area of turbulence that caused a nose-down pitch excursion of enough force to move a rather heavy suitcase from the baggage area far up into the cabin. This adds emphasis to the thunderstorm admonition about securing anything that you don't want to hit you in the head.

While pitch control might seem a problem, maintaining a wings-level attitude (and thus an approximate heading) is what gets the airplane through the storm and keeps it out of trouble. In light airplanes, a lateral upset would almost always precede a true loss of pitch control that would in turn result in an increase in airspeed to a dangerous level. In heavier airplanes, pitch upsets might be a more likely occurrence. While the airspeed of the heavy would react more slowly, because of the airplane's greater mass, it would recover more slowly. Thus if the airplane were stalled, it might stay that way for a while. In severe turbulence in a light airplane, you might in a time of high positive g-loading occasionally get the stall warning at maneuvering speed, accompanied by pitch down. But it's momentary.

It is possible to practice instrument flying in turbulence, and every serious instrument pilot should do so. While windy days don't duplicate thunderstorm turbulence, they do provide similar distractions.

TAKEOFF AND LANDING

Airport operations in the vicinity of thunderstorms are pretty well covered by logic, with wind a big factor. The characteristics of wind flow around thunderstorms put the pilot in a dilemma. If the wind is expected to shift or increase in velocity, the best way to take off would be in the direction of the expected shift. For example, if the wind is relatively calm and a thunderstorm is approaching from the southwest, as the storm moves closer the outflow from it is likely to result in a southwest wind. So take off to the southwest. Okay, but that is toward the storm. And if the outflow reaches the airport before takeoff, then the shear turbulence will be a big factor for climb-out. The best procedure is to delay the takeoff until the thunderstorm has passed. If you are headed east, that might mean having to fly back through the thunderstorm area after takeoff, but at least in that case you have the flexibility of choosing an altitude and a path. Starting from the ground, you have to use the runway, and until it has climbed for a few minutes the airplane is inevitably vulnerable because of low altitude and airspeed. The air carriers more or less solved their en route thunderstorm problems rather soon after the advent of jet transports. But on the basis of the evidence, it took them a lot longer to learn that the key to the thunderstorm question in the approach and departure phases of flight is abstinence.

Abstinence. That is why any discussion of thunderstorm flying needs quote marks around the word "flying." The subject is related more to flying around thunderstorms than to flying in them.

EPILOGUE

There are no pat answers to the thunderstorm question. If we totally stopped all operations when thunderstorms were about, the air transportation system would become a shambles on many days. Likewise if the forecast of thunderstorms or the lack of such a forecast is allowed to influence operations or pilot thinking, utility or safety will suffer. If we don't fly because they are forecast, we won't fly much. If we fly with a closed mind on the subject when they aren't forecast, we'll someday find an unpleasant surprise. The fact is that the forecasts we use are based on old information and are approximate at best. The key is in the general weather situation and in what is happening *right now.* And if a pilot is realistic, the cockpit is probably the best place in the world to observe weather and make judgments on the quality of flying weather.

And as pilots move around with a good store of

weather wisdom and perhaps an item of on-board weather avoidance gear, they soon learn that the thunderstorm question is very manageable. It's a matter of recognizing that, when pitted against one another, pilots and storms each have advantages. The winner of the game is the player who makes the most of his advantages.

The storm is by far the larger and the tougher. Even the energy expenditure of a Boeing 747 is negligible when compared with the energy found in one mature thunderstorm cell. True, the thunderstorm's energy is expended over a relatively large area, and more forcefully in spots than in others. But predicting the exact locations of strengths and weaknesses with any degree of accuracy is impossible. Destructive potential might be in any part of the storm. So if the first consideration of the contest is brute strength, the storm gets the first point.

But the pilot also has strong advantages. He (hopefully) has patience, vision, and a brain, and the airplane is far far faster than the storm. Intelligence and quickness count for a lot in any contest. The storm can't see, track, or catch the airplane. Actually it's really no, or little, contest. The storm hardly has a chance if the pilot plays the airplane's advantages. And indeed the airplane wins many more than it loses. Even when a pilot errs and winds up in a storm, the airplane usually wins after a tussle that leaves it banged around and quite wet. But that's not the object—that's a brush with possible disaster and an escape tainted by that unreliable aeronautical commodity, luck.

The real points are scored by pilots who both avoid

the turbulence of the storm and usually get where they are going reasonably close to schedule. That's the true contest and is where the fine points of thunderstorm "flying" come to play. The pilot who wins here is the one who will look and poke and probe in search of a relatively smooth path through an area of thunderstorms. But he's always ready to cut and run, or like a quarterback who hears hoofbeats, to fall on the ball. The winner knows that, in the end, the airplane has the ultimate advantage over the thunderstorm: staying power. That's where patience comes into play. The airplane can afford to wait. Properly flown and maintained, an airplane will last a long, long time. Thunderstorms only last a little while. The airplane that flies around them at a respectable distance, or waits on the ground for the time it takes for a storm to quit menacing the departure area, or diverts to an alternate airport to wait for activity to clear the destination, is the airplane that will be back to fly again another day.

INDEX